How to Guide to Choosing Laboratory Instruments

Money saving tips and guidance for environmental labs

William Lipps

Copyright © 2019 William Lipps

ISBN: 9781671161771

All rights reserved.

Contents

Forward 8

Introduction 9

    The instrument demo 15

    Buying used instruments 17

Laboratory Instruments 23

    The balance 24

        Suggestions for maintaining and selecting a balance 25

    pH and ISE Meters 28

        Suggestions for selecting a pH or ISE meter 29

    Automatic Titrators 33

        Suggestions on selecting an automated titrator 35

    TOC Analyzers 36

- Combustion TOC 38
- Wet Chemical Oxidation TOC 41
- Choosing a TOC analyzer 42
- Molecular Absorption 49
- Automated Chemistry Analyzers 58
- Atomic Absorption Spectrophotometry 62
  - Flame AA 62
  - Graphite Furnace AA 64
  - Hydride Generation and Cold Vapor AA .. 66
  - Suggestions on selecting AA instruments 69
- Inductively Coupled Plasma – Atomic Emission Spectroscopy 71
  - Suggestions on selecting ICP-AES instruments 103
- Inductively Coupled Plasma – Mass Spectrometry 104
  - Suggestions on selecting ICP-MS instruments 114
- Chromatography 117

- Gas Chromatography (GC) 120
- Gas Chromatography Mass Spectrometry (GCMS) 122
  - Suggestions on selecting a GCMS 131
  - Features to compare on GCMS Systems: 135
  - Triple Quadrupole GCMS 137
- Ion Chromatography 149
  - Suggestions on selection an ion chromatograph 154
- High Performance Liquid Chromatography (HPLC) 156
- Liquid Chromatography Triple Quadrupole Mass Spectrometry (LCMSMS) 166
  - What Are the Advantages of LC-MS? 176
  - Suggestions for selecting an LCMS, or LCMSMS 177
- Summary 179
- About the Author 181

Forward

The conclusions presented here are from my experiences as a lab owner, employee, and working as an instrument manufacturer. I grew up in the commercial laboratory business, before there was such a thing as an environmental lab. Our business was our bread and butter. If we could not run samples, we did not eat. Therefore, it was very important that the laboratory equipment be working. Most of my conclusions presented here are observations based on this past, and from talking with people like you at laboratories all over the world. Please note that unless otherwise stated, all graphics were obtained from Shimadzu.

Introduction

We buy laboratory instruments because we do environmental testing. We do this testing because the law requires it. This requirement by law is fundamentally different from other types of testing that measure the quality of a product. We do environmental testing because we have to, and we have to do the testing by methods prescribed by regulatory agencies.

Environmental regulations specify the methods. For example, the Safe Drinking Water Act defines certain pollutants in drinking water and requires measurement by certain methods.

Similarly, the Clean Water Act defines a set of methods to follow when analyzing wastewater.

Although SW846 methods are mostly guidance methods, RCRA still lists methods and you are still required to run them to comply with solid waste regulations or contract requirements.

Why do laboratories run these methods? Obviously, a commercial lab runs samples in an effort to make a profit. A municipality may run samples for process control, or they may run them to prove that their stuff is below limits established by the EPA or another regulatory agency.

All methods require some kind of instrument. Even the simplest of laboratory methods will require a balance or some type of device.

So, how do you decide what you need? To do so, do not just look at your needs today. You need to think of next month, next year, and five years down the road. Do not replace what you have with what you have. Improve! Find the best fit. Find a manufacturer willing to consult with you. Let the manufacturer help you to decide what is best for you. Even if what is best for you is not their own product.

When choosing a laboratory instrument remember you are buying it to run a method. Your methods are prescriptions and have limited flexibility for modification. The instrument is, or contains, the detector. The detector is usually NOT included as one of the things that you can modify in an EPA method. Detector definitions can be very specific. You must use the same detector technology used during validation of the approved method. Make sure that whatever you buy is the same, or an allowed, technology.

Forget bid specifications, at least at first. Do not get hung up on that. You want an instrument that will enable you to run the EPA approved method with very little disruption. Read the methods (caution some are very old) Search the internet (although that can be very confusing). Decide what methods you are going to run in what matrices and for what "regulatory acts". Decide your required detection limits for each analyte based on the method requirements, permits, and historical or competitive data. Do not request obscenely lower detection limits than you will ever need.

Any purchase of a laboratory instrument must fit the needs of the method. Buy what you need, now and in the future, but do not buy more than you need. Purchase the instrument at the right price. A price that you can afford, but also one that the vendor can afford. You do not want the vendor to lose money. You do not want to buy an instrument from a company that goes out of business a few years later.

The instrument MUST be able to run the method. That is the most important thing. Do not get hung up on specifications. These features are ways manufacturers differentiate their products. Usually these features are ways individual manufacturers use to do the same thing that another feature, on a different product may do. Manufacturers create bid specifications and you may use them to lock out another company if you do not want that product. Fine, it is your choice. Just be careful that you do not box yourself in and prevent yourself from using newer technology that might get you better turnaround, less maintenance, or lower detection limits.

Always remember that a poorly selected instrument will affect many things. It will affect your results, your data, your productivity, and maybe even your job.

I know that it is the safe bet to go with the market leader, or the one that everyone else has. Just be sure that if you decide to do so you are not limiting yourself. The reason there is competition is that other companies believe their instruments can run the same tests. You never know. That different brand may be a better fit in your lab. The service and support may be better. It may be easier to maintain. It may improve detection limits or productivity. Be careful when locking others out just to stick with status quo.

Do your research. Go on-line, and to trade shows. Talk to specialists at the booths. Compile your list of needs. For example, "I need a GCMS with a purge and trap to analyze water samples by EPA Method 524.2". Now you know exactly what you need. Methods define sample preparation, analytes, and detection limits.

Now that you know what you need, you need to decide which one. Let the manufacturer sales staff do your work for you. Let them tell you what instrument in their portfolio is best for your chosen analysis. Let them provide you with the specifications and show you data demonstrating that method on that instrument. Do this with each manufacturer.

The instrument demo

There is no such thing as a perfect demo. Lab instruments are complex. You do not do demos on pH meters. Before you ask a company to bring an instrument in for a demo, work with them and decide what success is. If you have already decided on manufacturer x it is not fair to manufacturer y for you to force them to spend money and come do a demo just because. In addition, manufacturer y's instrument may not operate just like manufacturer x. Allow manufacturer y to bring his own reagents and calibrate according to "manufacturer instructions".

Many times, instead of a demo you send samples to the manufacturer. If you do this, buy a reference material and send that. If you split an unknown sample between manufacturers, how do you know whose result is the correct one. If you send a complex matrix, prepare them in advance. They do not know the history of your samples.

If you send a complex sample, send enough of it so the vendor's lab can do testing on their own. Then, if they come up with a valid solution to your tough matrices let them recommend it, even if it is not the answer you wanted.

Buying used instruments

Buying a used instrument can appear, on the books anyway, to save a lot of money. However, does it?

Before you buy a used instrument, decide what you need. If you are simply getting another of the same to avoid re-writing a SOP, the search is simple.

Before you pay, find out whether it is re-furbished, reconditioned, or sold "as is". Make sure it has a warranty. Most manufacturers will not support older instruments.

- Refurbished means some qualified technician certified it to be "like new", Find out if that is true. Find out what was replaced, and why. How did they test it? Where did they test it?
- Reconditioned instruments are old demo instruments, or instruments that were returned for some reason. Ask the vendor for testing data.

Make sure they made sure it is still working.
- Condition is unknown to the seller when you buy an instrument "as is"

Before you buy used, ask yourself "is it really worth it"? Why did the other person sell it? In addition, remember Moore's law (Figure 1):

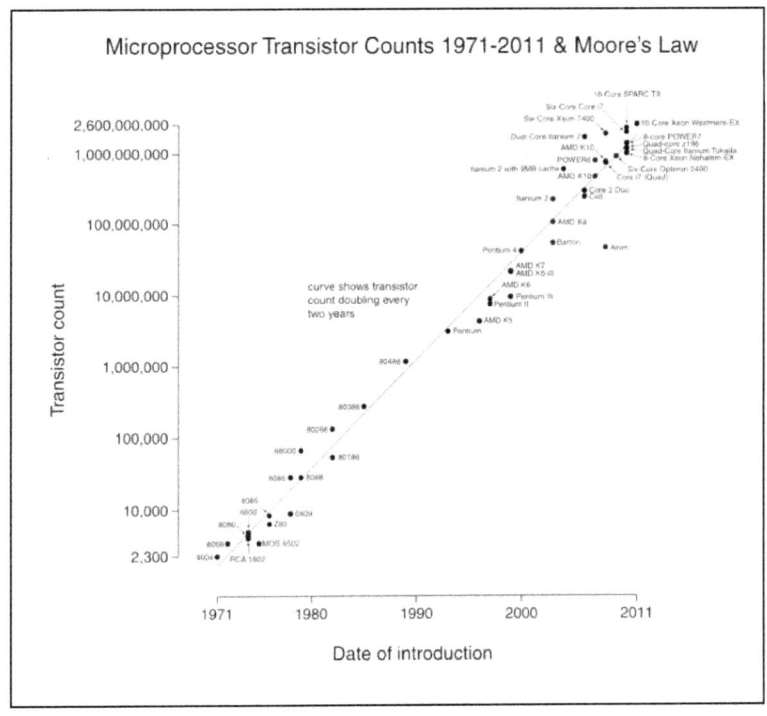

Figure 1, Illustration of Moore's Law[1]

---

[1] https://en.wikipedia.org/wiki/Transistor_count accessed February 2, 2017

According to "Moore's Law", the transistor count of integrated circuits doubles about every two years. This results in rapid development of faster computers and new operating systems.

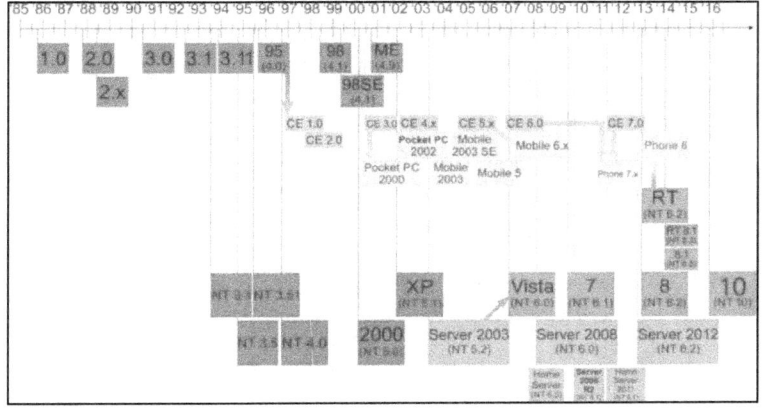

Figure 2, Windows timeline[2]

---

[2] https://en.wikipedia.org/wiki/Timeline_of_Microsoft_Windows, accessed February 2, 2017

Figure 2 shows how rapidly Windows updates their software.

When Windows forces you to update computers, your older instruments may no longer work. You will have to have a storehouse of old computers just to keep running.

However, detection limits are not staying the same. They are decreasing. To keep up with the Jones you will have to invest in newer instruments capable of achieving the lower detection limits.

Manufacturers are constantly improving speed, sensitivity. Manufacturers write new software on newer operating systems. Eventually, your older instrument will be incompatible.

My suggestion is to buy new. Do not risk your reputation on old instruments that require outdated operating systems that you cannot even buy computers for anymore.

Laboratory Instruments

The balance

The most important laboratory instrument in your lab is the balance. The degree of accuracy in your lab is proportional to the accuracy of all the standards and reagents you prepare. The balance should be the most cared for and protected piece of equipment in your lab. However, may often overlook proper care and placement of their balance.

There are more uses for a balance than the preparation of reagents and standards. Final mass measurements for several important environmental tests, including those listed in Table 1, are made on the analytical balance. If the balance is not operating properly, your results for these tests could be wrong.

Table 1, Methods that rely on the analytical balance for final mass measurement

Method	Analyte
SM 2540B	Total Solids (TS)
SM 2540C	Total Dissolved Solids (TDS)
SM 2540D	Total Suspended Solids (TSS)

Suggestions for maintaining and selecting a balance

Mount the analytical balance on a heavy, shockproof table with a large surface area and a drawer to keep balance accessories. Check the balance level daily. Keep balances away from doors, air conditioner vents, direct sunlight, and laboratory traffic. Equilibrate the balance with room temperature before use. If your A/C shuts down when you are not there, wait until the lab temperature is stable before using the balance. Cleanup spills immediately.

Balances come as top loading, analytical, micro-analytical, etc. Each with its own use. At minimum, you will need a good top loading balance and an analytical balance. The top loader you will use when preparing reagents, and the analytical you will use for final mass measurements of TSS, TDS etcetera and for weighing material in the preparation of calibration standards.

Fortunately, all analytical balances of the 200 gram or so capacity that are suitable for environmental labs have about the same specifications with reference to sensitivity, precision, convenience, and price. Since it is safe to assume that about any balance you buy is suitable, the selection can be made on extra features and perhaps availability of service. One feature is "self – calibration" with temperature fluctuations in the lab. If your lab temperature is fairly well controlled, you may not need this. Another feature automatically calibrates with internal weights. Simply zero, then calibrate. Even so, you still need to verify the calibration with an external, NIST traceable weight.

Other options are available and will depend on your preferences. Link to a LIMS or software that will minimize transcription errors or automatically calculate is nice. Printers can be used to cross check manual notebook or worksheet entries.

pH and ISE Meters

A basic meter consists of a voltage source, amplifier, and scale or digital readout. Additional features can produce varying performance between models and manufacturers. Some models may feature expanded scales, solid-state circuitry, and temperature and slope adjustment. Other features facilitate the use of ion selective electrodes, recorder output, or interfacing with a LIMS.

In routine pH measurement, the glass electrode is the indicator electrode and a calomel electrode serves as the reference. Glass electrodes have a very fast response time in buffered solutions; however, not all samples are buffered. Fortunately, as with the laboratory balance, pH and ISE meters have been standardized to the point that if, you follow the manufacturer's instructions to calibrate; all meters should get close to the same result on just about any sample.

Suggestions for selecting a pH or ISE meter

The accuracy of the meter does matter. For example, a difference in 5 millivolts for pH is a difference in 0.1 pH units. It is important to ensure your meter is functioning properly, and your reference electrode is good. For a monovalent ion, a 1-millivolt error results in 4% error in the result. For these reasons, I am suggesting that you get a meter with +/- 0.01 mV resolution and a pH meter with a minimum resolution of 0.01 SU.

Other things to look for are water and impact resistant buttons. Although you may say "but I will only use this in the lab", pH is a messy test and the odds are you will spill sample or buffer on the meter a few times. Unless you are very happy with your particular manufacturer and are very confident the manufacturer will never go away or run out of stock, look for universal connections such as the BNC. This will allow you to purchase new electrode from whoever and whenever you need them. Look for meters that allow more than a two-point calibration. This is especially helpful for ISE and when measuring pH below and above seven.

Do not get an analog instrument, if they are even available anymore. They are too hard to read. Get a digital readout. However, remember, that last digit is always questionable. If you really want resolution down to 0.01 units then get a meter capable of resolution to 0.002 units. Make sure your meter enables you to check the slope. Older meters had you adjust the slope, but with automatic calibration, slope adjustment is done for you. Nevertheless, methods and your QA/QC program likely require some sort of slope verification.

The electrode matters, and your reference electrode really, really matters. There are hundreds of electrodes making selection very confusing. I can only speak to my experience here, but a low cost glass pH electrode surrounded by a ceramic frit does yourself no favors. Imagine my dismay at missing my high pH WP samples. I remedied this by spending a little more on an electrode that enabled me to refresh the liquid junction. If the liquid junction fails, the reading is wrong.

Ion selective electrodes are more expensive than glass pH electrodes. For this reason, I recommend getting a separate reference electrode. An exception, of course, is the dissolved oxygen probe. To contradict myself even more, I have had plenty of luck using the fluoride combination electrode.

Table 2 lists methods that use ion selective electrodes for measuring the analyte. With the exception of pH, all of these electrodes measure the activity of the analyte in a buffered solution. Recall that for pH the standards are buffered but the samples are not. Take care not to damage the indicating crystal. Ammonia and nitrate membranes need routine replacement. The ionic strength adjustment solutions used in the methods contain complexing reagents that minimize interferences. Chloride, for example, interferes with the nitrate measurement. Be sure and follow the manufacturer's instructions to compensate for these interferences.

Table 2, Methods that use ISE or pH

Method	Analyte	Sensor Material
SM 4500-H$^+$	pH	glass
SM 4500-CN$^-$ F	Cyanide	crystal
SM 4500-CL$^-$ D	Chloride	crystal
SM 4500-F$^-$ C	Fluoride	crystal
SM 4500-NH3 D or E	Ammonia	membrane
SM 4500-NO3$^-$ D	Nitrate	membrane
SM 4500-S^{-2} G	Sulfide	crystal

Automatic Titrators

Automatic titrators use pH electrodes, ion selective electrodes, coulometry, or simple photometers to detect the endpoint of a titration. They may or may not; use auto-samplers aka sample changers. The titrator determines the amount of analyte in a sample when standardized reagent is added.

The standardized reagent is added with a burette, but much more accurately and reproducibly than is possible manually. Systems can be dedicated by test or purchased /configured to run multiple tests. A mono burette system is adequate for most purposes. Some manufacturers configure them so that you can run fluoride by ISE, determine pH and then titrate alkalinity all on one system.

Table 3 lists methods for which you can use an automated titrator. While most of these parameters can be determined in other ways, there is no alternative, other than titration, for alkalinity.

Table 3, Methods that allow automatic titrators

Method	Analyte
SM 2320 B	Alkalinity
SM 2340 C	Hardness
SM 4500-CL⁻ D, E, F	Chloride
SM 4500 CN D	Cyanide
SM 4500-NH3 C	Ammonia

Suggestions on selecting an automated titrator

When looking for a titrator check if probes have proprietary connections, the cost of replacement electrodes and software updates. Look for availability of on-site service, whether it can be services locally, or how long it would take to be repaired. Most people that buy these are looking for a dedicated, easy to use system.

TOC Analyzers

Total organic carbon analyzers measure the amount of organic, inorganic, or total carbon in water or soil samples. TOC is an important indicator of disinfection byproducts and the byproduct rule requires drinking water facilities to measure TOC removal. TOC also correlates with BOD or COD in many matrices and can be used as a surrogate for those tests.

TOC is analyzed by either removing all the inorganic carbon and measuring NPOC, or it can be calculated by subtracting inorganic carbon from total carbon. The subtraction method is prone to error and most people actually just measure NPOC.

The primary differences between TOC analyzers is the mode of detection, and the oxidation technique. Most TOC analyzers convert all carbon to carbon-dioxide and measure the CO_2 with a CO_2 specific detector. These are either IR detection, or gas diffusion with conductivity detection. IR detection requires CO_2 be swept by a carrier gas through the detector. It is more specific than membrane conductivity but it needs a carrier gas. Membrane conductivity diffuses CO_2 through a membrane into an absorber solution. Any change in conductivity of the absorber solution measures CO_2.

EPA approved methods for TOC oxidize organic compounds using either high temperature combustion or a wet chemical oxidation. High temperature combustion may occur by collision of gaseous phase sample with oxygen or by collision with oxygen on the surface of a catalyst. The catalyst-assisted oxidation allows lower temperatures to be used increasing combustion tube life. Most wet chemistry oxidation is using the persulfate ion. Persulfate oxidations are catalyzed by heat, UV irradiation, or a combination of the two. Another oxidation technique, recently approved as an ATP for drinking water, uses alkaline hydroxyl radicals. This technique overcomes chloride interference typical of persulfate oxidation.

Combustion TOC

Combustion is a series of complex reactions between fuel and oxygen similar to what happens in this flame. Inside the TOC combustion tube, the sample, or fuel, heats to gaseous phase and collides with oxygen molecules. These collisions result in CO_2 and H_2O. The efficiency and capacity of the reaction increases with oxygen, most combustion analyzers need pure oxygen for complete recovery.

In the TOC analyzer, the sample is injected into a stream of oxygen that flows through a heated tube and then through the IR detector. You need as many collisions as possible between the organic compounds and the oxygen. The higher the temperature of the combustion tube the more collisions. Thus, combustion analyzers are operated at temperatures near 1000 °C.

Catalytic Combustion oxidizes organic compounds to CO_2 and H_2O at lower temperatures

The carrier gas containing oxygen passes over a hot metal catalyst. The oxygen molecules split to oxygen atoms that coat the surface of the catalyst. Sample, or fuel, is injected and splits into atoms on the surface of the catalyst. In Figure 3 example, the fuel is methane. A carbon from the methane collides with an oxygen atom to form carbon monoxide. Because of its affinity for the catalyst the CO molecule stays put until it collides with another oxygen to form CO_2. The CO_2 flies off into the carrier gas. The hydrogens from the methane also collide with oxygen to form OH^- first then H_2O, which also enters the carrier gas. This process continues as the fuel percolates through the catalyst bed. Once there is no more fuel, the catalyst recoats with oxygen from the carrier gas.

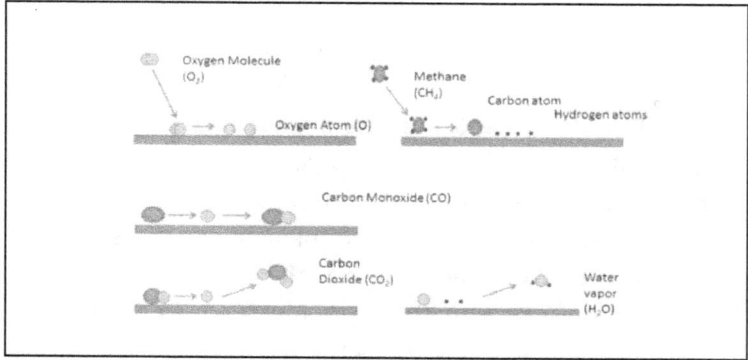

Figure 3, Example of catalytic combustion of methane

The mode, or mechanism, of collision is different with catalytic oxidation as compared to combustion. In combustion, the collisions occur in the gaseous phase requiring very high temperatures to induce enough collisions in the shortest time possible. The carrier gas needs to be pure oxygen to ensure adequate collisions. Because the catalytic oxidation takes place on the surface of the catalyst, the temperature can be lowered. The lower temperature is below the melting point of sodium chloride decreasing devitrification and increasing the life of the combustion tube.

Wet Chemical Oxidation TOC

Chemical oxidation reacts an oxidizing chemical with organic compounds to CO_2 and H_2O.

Under acid conditions, persulfate converts into a sulfate radical, which is a very strong oxidizer. Unlike catalytic combustion, that takes place between a gaseous and solid surface, persulfate reactions take place in solution. Persulfate is not so good an oxidizer on its own, but its oxidizing power increases with heat or upon exposure to UV irradiation. The number of collisions between the oxidizer and the sample increases with temperature, unfortunately, the persulfate rapidly decomposes at high heat. In addition, other constituents of the sample matrix, particularly high chloride, react with persulfate preventing it from oxidizing the carbon.

Choosing a TOC analyzer

The two main oxidation techniques for TOC analyzers kind of play out like this. Combustion, in its true form, requires an oxygen carrier gas, has higher detection limits. This is what you find in solids TOC analyzers. Catalytic oxidation TOC analyzers operate at lower temperatures, preferably below 800 °C (the boiling point of NaCL). This way they can tolerate high salt content. They have higher detection limits than persulfate analyzers because you cannot inject as much sample. Persulfate analyzers have lower detection limits but are more susceptible to interferences from Chloride. My suggestion is that for quantitation below 500 ppb get a persulfate analyzer. Between 500 ppb and about 2 ppm, it is a tossup, and above 2 ppm go with catalytic combustion. Why? Its faster, uses fewer reagents, and has a better oxidation efficiency (Table 4).

Table 4, Comparison of TOC oxidation technique

Technique	Advantages	Disadvantages
Combustion (HTC)	High efficiency Soil, coal, solids	High detection limits Hazardous temperatures
Catalytic Combustion (HTCC)	High Efficiency Lower temperature than HTC Tolerates High Salt	Higher detection limit than PO Memory effects of catalyst Catalyst poisoning
Chemical Oxidation (PO)	Low Detection Limits No memory effects	Chloride interferes Lower oxidation efficiency

The use of these two oxidation techniques for environmental samples is split almost 50/50 (Table 5). Catalytic combustion has a slightly higher usage. As mentioned before, they are good in samples with high matrices, such as wastewater, and seawater. People usually purchase persulfate analyzers when they want to measure TOC below 500 ppb. The persulfate analyzers have better detection limits because you can inject more sample. However, because it is a chemical oxidation the efficiency may not be as high. It is a definite tradeoff. Some manufacturers claim their catalytic combustion analyzers can accurately detect TOC below 500 ppb even into the 10-ppb range.

Table 5, Use of TOC oxidation techniques

Method	Oxidation Technique	% Users
SM 5310 B	Catalytic Combustion	44
SM 5310 C	Chemical	35

Measuring low concentrations of TOC means you need certified "low TOC" vials. Table 7 is an example of a test I did using a persulfate method measuring low TOC in regular VOA vials, often used as sample collection for TOC analyzers. Notice the two high values that caused the MDL to be high. These high values are due to contamination in the vials.

Table 7, TOC MDL using regular VOA vials

Run 1	57
Run 2	46
Run 3	45
Run 4	98
Run 5	49
Run 6	80
Run 7	52
Standard Deviation	20.2
MDL	63 ppb *Ouch!!*

Table 8 is a repeat of the same test using low TOC vials. Notice a much better MDL. However, also note that these were not spiked samples. This test uses laboratory grade water. The TOC detected is a combination of the TOC in the laboratory reagent water, and the TOC contamination of the low TOC certified vials. There is no such thing as TOC free water and no such thing as TOC free vials. Keep this in mind when evaluating TOC analyzers; a TOC analyzer measures TOC concentrations in the sample bottles, your lab water, and contamination from all sample manipulation. This may raise your detection limits to higher than those listed in the manufacturer's specifications.

Table 8, TOC MDL using low TOC vials

Run 1	33
Run 2	28
Run 3	22
Run 4	21
Run 5	31
Run 6	28
Run 7	26
Standard Deviation	4.397
MDL	14 ppb

Molecular Absorption

The UV-visible spectrophotometer is a workhorse in almost every environmental lab. The versatility of these instruments and the wide variety of uses have resulted in a variety of designs and different price ranges. The essential parts of a spectrophotometer are a source of light, a monochromator, a cuvette to hold the sample, and a photodetector. The type of monochromator and photodetector are prime differentiators in design from one instrument to another and between manufacturers.

The filter photometer (Figure 4) is the simplest, and cheapest. These use a filter that only allows one wavelength of light to pass. They are essentially method specific, although they can run several common methods, such as phosphate and ammonia, using a single wavelength. Field portable meters, such as chlorine meters, are filter photometers. Many of the lower cost spectrometers used with manufacturers test kits are filter photometers.

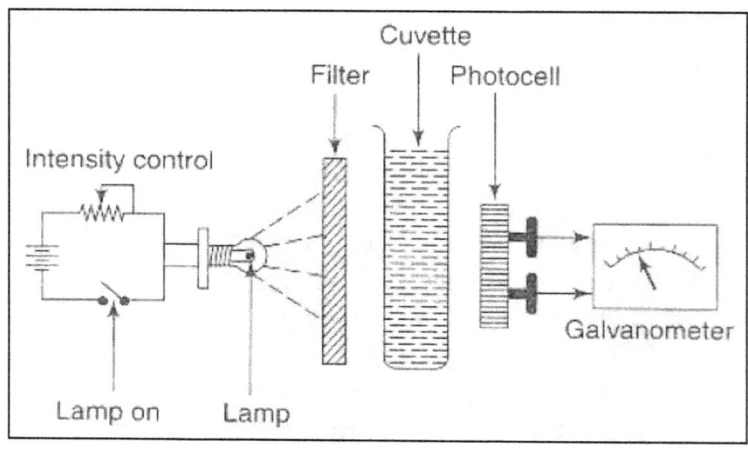

Figure 4, Filter photometer[3]

A filter photometer is usually restricted to visible wavelengths. They are not complex instruments and most likely have pre-stored calibration curves for use with prepackaged reagents. They are suitable for almost every

---

[3] http://www.myclassbook.org/single-beam-filter-photometer/, accessed February 6, 2017

environmental test, but you cannot expand to tests for which there is no filter. The advantage is that they are low cost, highly reproducible wavelengths, simple to operate and very useful for routine tests.

A prism or grating spectrometer (Figure 5) disperses light and then passes a single wavelength through the sample. These photometers often have an adjustable slit. You can make the band pass wider to increase the sensitivity; however, linearity at the upper end of a calibration may be less.

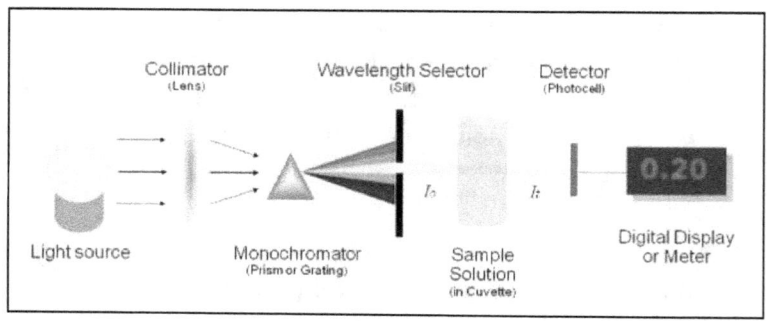

Figure 5, Prism Photometer[4]

Grating or prism photometers are also generally restricted to visible light. However, by spending a little extra compared to a filter photometer you now have more wavelengths so that you can add more tests later. The adjustable slit allows you to gain a bit of extra sensitivity, or

---

[4] http://www.slideshare.net/nasirnazeer5/spectrophoto-meter-31242060, accessed February 6, 2017

you can narrow the slit and possibly increase the maximum range.

The most expensive, upper end are UV-Visible spectrophotometers (Figure 6). These pass full spectra through the sample enabling acquisition of a complete absorbance spectrum.

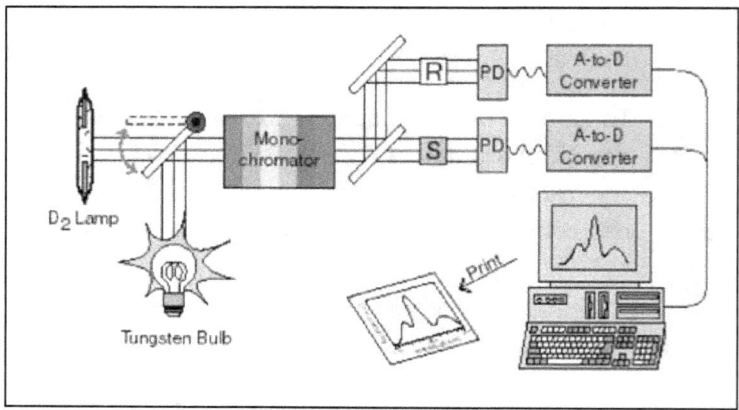

Figure 6, Scanning UV-Visible spectrophotometer[5]

Full spectral analysis enables analysts to select the peak maximum or perhaps measure off peak to increase the linear range, measurements in the UV range and the entire visible spectra making it a very versatile and

---

[5] http://www.rohs-cmet.in/content/uv-visible-spectrophotometer, accessed February 6, 2017

useful investment if you plan to do anything more than a set of routine tests.

Tips for using a photometer or spectrophotometer

Anything that absorbs light will be measured. Filter samples. Remove air bubbles. Clean fingerprints off the cell. For the greatest accuracy, use a single quartz cell. If not a single quartz, then used matched cells.

If you use a sipper with a flow cell, however be aware that flow cells can get dirty. Keep in mind that lenses, mirrors, and gratings fog over time; an old spectrometer may not perform as well as a new one. Keep the instrument in an acid free area with low humidity and minimal vibration.

Tips for selecting a photometer or spectrophotometer

Table 8 lists examples of visible and UV absorbance methods. With the exception of UV-254, all are possible on a filter photometer. Your choice of photometer or spectrophotometer has much to do with use. If you will simply run the same tests repeatedly, then a filter photometer may be sufficient. If you want to add new tests later, then a grating or prism visible spectrometer may be the instrument of choice. More sophisticated tests that need multiple wavelengths or measure in the UV range need a spectrophotometer. The more expensive spectrophotometers will likely be able to interface readily with a LIMS, or operate with GALP compliant software.

Table 8, UV-Visible Methods

Method	Analyte	Wavelength (nm)
SM 4500-F D	Fluoride	570
SM 4500-CN⁻ E	Cyanide	578
SM 4500-CL G	Chlorine	515
SM 4500-NH3 D	Ammonia	630
SM 4500-P E	Phosphorus	880 (or 660)
SM 3500-Cr D	Chromium VI	540
SM 3500-Fe D	Iron (II)	510
SM 5910 B	UV-254	253.7

Automated Chemistry Analyzers

Automated chemistry analyzers automate the manual color reagent additions that you must do when doing colorimetric environmental tests. The advantage of automation is that reagent addition and measurement time is very precisely reproducible. Automated analyzers use less reagent volumes and generate less waste than manual methods. Thus, automated wet chemistry analyzers can improve your data, decrease labor and reagent costs.

Almost any colorimetric test can be automated. The most common automated analyzers are segmented flow analyzers, flow injection analyzers, and discrete analyzers. A discrete analyzer uses less reagent than the flow chemistries; however, discrete analyzers are more complex.

Tips for choosing an automated chemistry analyzer

The discrete analyzer has a fixed sample path. One manufacturer gives you a choice of a couple of different path lengths. Flow analyzers can have a range of user changeable path lengths. Flow analyzers usually get better detection limits. Discrete analyzers are highly computerized. Keep service in mind if buying a discrete analyzer. A flow analyzer has almost no easily broken hardware requiring much less service. The flow analyzer does require lots of analyst manipulation in changing pump tubes and connecting tubing and coils. Discrete analyzers consume much less reagent, but require either disposable, or somewhat disposable cuvettes. Flow analyzers require pump tubes and replacement of cartridge tubing.

Parameter	Written for	Method
Ammonia Nitrogen	SFA	EPA 350.1
Nitrate Nitrite Nitrogen	SFA	EPA 353.2
Phosphorus	SFA	EPA 365.1
Available Cyanide	FIA	ASTM D6888
Total Cyanide	SFIA	ASTM D7511
Nitrate Nitrite Nitrogen	Discrete	ASTM D7781

Table 9, Automated Chemistry analyzer methods

Table 9 lists methods that require flow or discrete analyzers. CFR 40 part 136.6 allows interchange between manual, flow, and discrete methods.

Atomic Absorption Spectrophotometry

Atomic absorption spectrophotometers measure concentrations of metals and semi-metals in low ppb to ppm levels. Free atoms absorb element specific radiation emitted by a lamp. Atomization can be in a flame, a graphite furnace, a heated tube, or generation of vapor.

Flame AA

Flame AA (Figure 7) detection limits range from about 1 to 100 parts per billion depending on the element. The liquid sample is aspirated into a nebulizer and smaller droplets are drawn into a flame. Elements atomize in the flame. Element specific light, passing through the flame is absorbed by the free atoms in proportion to their concentration.

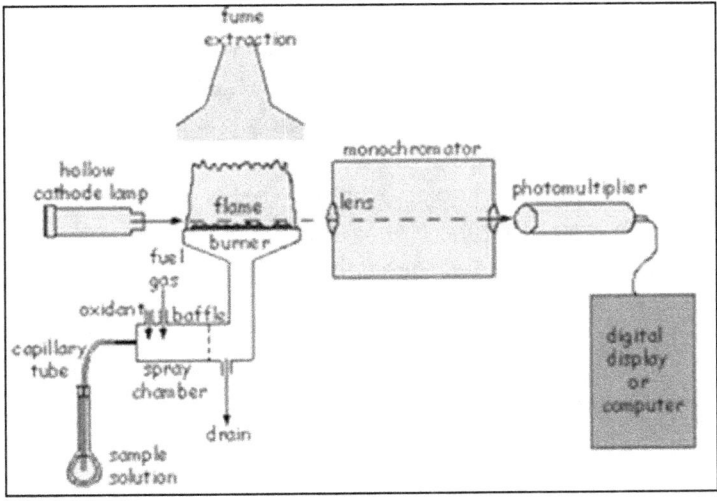

Figure 7, Schematic of Flame AA[6]

---

[6] http://www.assignmentpoint.com/wp-content/uploads/2013/12/atomic-absorption-spectrometer.jpg, accessed February 12, 2017

Flame AA is not a complex instrument and very easy for non-technical people to run. If you only have a few elements, such as iron, manganese, and copper, to run then flame AA is an instrument of choice. Flame AA only runs one element at a time. You would run your iron, change the lamp, then your copper, and so forth.

Graphite Furnace AA

Graphite furnace AA (Figure 8) adds a modifier to the sample then injects it directly into a carbon tube. Current applied to each end of the tube heats it. First to dry the sample, then char, then atomization. The charring step removes interferences without volatilizing the analyte. The modifier is added to raise the charring temperature. Once charring is complete, the temperature is raised rapidly producing a signal as the atomized element of interest absorbs the element specific wavelength.

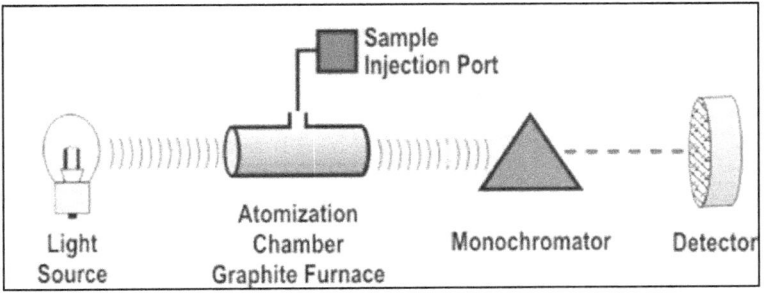

Figure 8, Schematic of Graphite Furnace AA[7]

Furnace AA has very low detection limits. It is slower than flame AA per reading and only measures one element at a time. Different elements may require different graphite tubes and different modifiers. Use furnace AA for elements with detection limits lower than possible by flame AA.

---

[7] http://image.slidesharecdn.com/graphitefurnaceatomicabsorptionspectroscopy-140719015607-phpapp02/95/graphite-furnace-atomic-absorption-spectroscopy-7-638.jpg?cb=1405734989, accessed February 12, 2017

Hydride Generation and Cold Vapor AA

Hydride generation (Figure 9) has an advantage over furnace AA in that it completely removes the analyte from the matrix for measurement. Use hydride AA for arsenic and selenium in high TDS samples. As with other AA methods, you only measure one element at a time. Unlike other AA methods that can be run from a single sample digestion. As and Se require separate preparation steps prior to hydride generation. This adds to the cost of analysis. In hydride generation, the As or Se are converted to hydrides by adding sodium borohydride. The gaseous hydride is swept, by a carrier gas, into a flame or heated quartz tube where it is atomized. The atoms absorb element specific light in proportion to concentration. Mercury is not converted to a hydride, but is measured in a similar fashion. Mercury also requires a separate digestion. Stannous chloride or sodium borohydride are added and mercury vapor is swept by a carrier gas into a non-heated tube. The mercury atoms absorb light in proportion to their concentration.

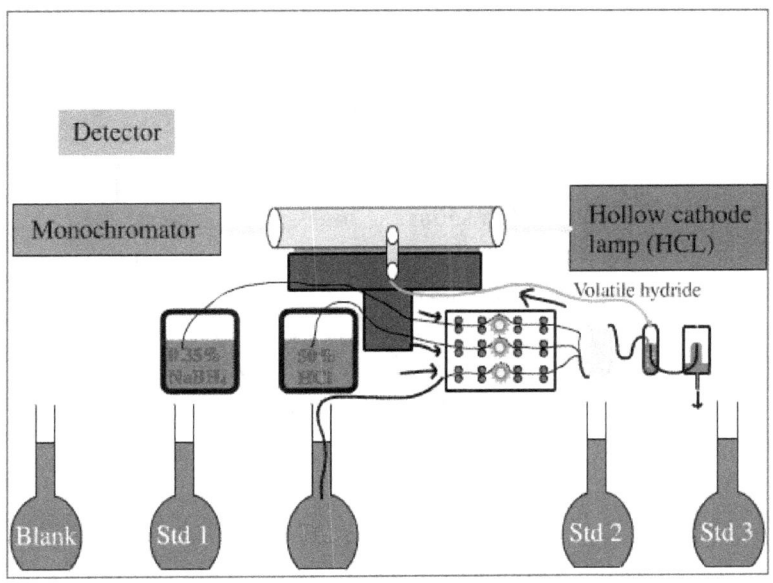

Figure 9, Schematic of Hydride AA[8]

---

[8] http://www.shsu.edu/~chm_tgc/primers/HGAAS_files/HGAAS3.gif, accessed February 12, 2017

Hydride and cold vapor also have low detection limits. The techniques serve to separate the analyte from the matrix minimizing interferences. Each requires a separate digestion and a separate analysis. Because of severe matrix interferences that can happen by furnace AA, you may need hydride generation of As and Se in some samples. Because of the separate digestions and analysis, hydride generation is a slow and labor-intensive technique.

Table 10 lists examples of methods that require the use of atomic absorption spectrophotometers. The Standard Methods for the Examination of Water and Wastewater methods are multi-element. EPA methods, with the exception of EPA 200.9 (a graphite furnace method) are element specific.

Table 10, Examples of AA methods

Method	Analyte	Atomization
SM 3111 A	Most metals	Flame
SM 3111 D	Refractory metals	Nitrous flame
SM 3113 B	Trace metals	furnace
SM 3114 B&C	As and Se	Hydride

Suggestions on selecting AA instruments

Double beam instruments can correct for drift that is possible as the intensity of the hollow cathode tube changes over time. Many instruments have a turret that holds and keeps warm, more than one lamp at a time. This shortens warm up time between elements. If you will need both flame AA and furnace AA look into how hard it is to switch back and forth.

Some instruments may do this automatically. Does the analyzer automatically light the flame via software? Can the software automatically choose the lamp and set up all operating conditions? Is it possible to have the software prepare calibrations from a single stock or dilute off scale samples?

Inductively Coupled Plasma – Atomic Emission Spectroscopy

Inductively Coupled Plasma – Atomic Emission Spectrometers (ICP-AES) are one of the most popular instruments in environmental labs because a single method/analyzer is capable of running almost every metal in a large number of samples per day. ICP spectrometers are very high throughput capable of multiple reportable results per run. ICP-AES is suitable for almost every element, excluding halogens and inert gases and are especially useful for refractory elements, such as silicon, aluminum, barium, etcetera, that perform poorly by flame AA. Samples are usually aqueous and are aspirated into a nebulizer. Method development is relatively easy; you can get by with analytical grade reagents. Once a method is set up, the instrument can be calibrated and operated by most lab personnel. The biggest drawback is lack of sensitivity for some elements, physical, and spectral interferences.

Figure 10, Elements amenable to ICP-AES[9]

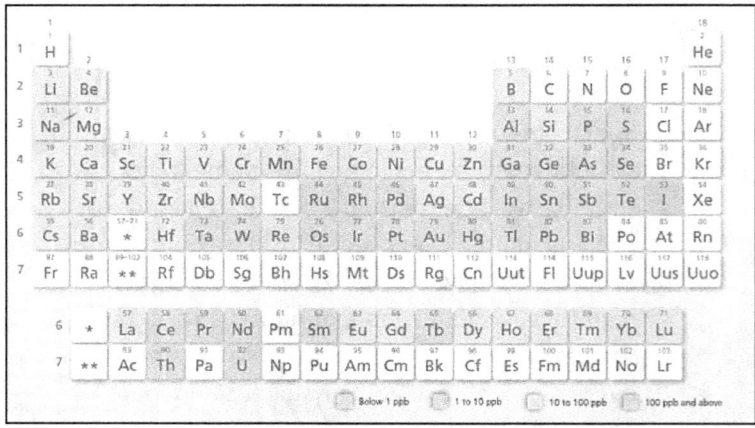

The colored elements in Figure 10 are those measured by ICP-AES. The various colors represent detection limits. Notice that elements, such as Ba, B, Al, and Si perform quite well by ICP.

9

http://www.ssi.shimadzu.com/products/product.cfm?product=icpe9800, accessed February 12, 2017

Figure 11, Schematic of a sequential ICP-AES[10]

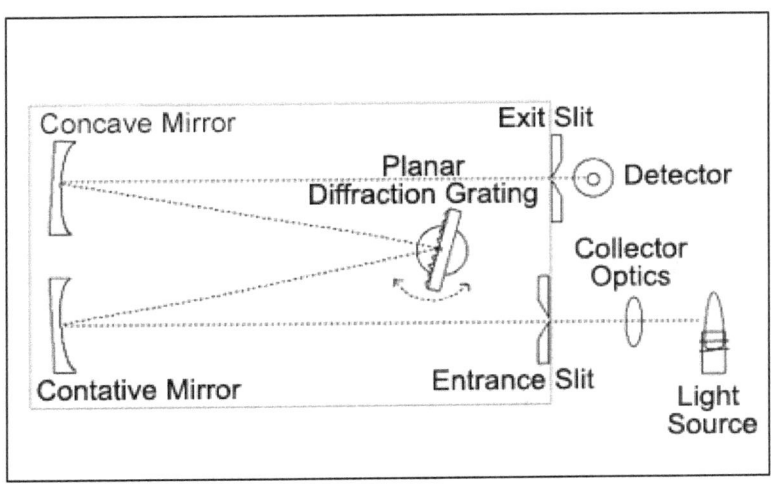

A sequential ICP (Figure 11) has a moving diffraction grating with a stationary photomultiplier detector. As the grating moves, each wavelength is measured. A sequential

---

[10] http://www.hitachi-hightech.com/image/global/products/science/tech/ana/icp/descriptions/icp-oes_01.gif, accessed February 24, 2017

analyzer has very high resolution with the ability to eliminate spectral interferences. They are also; however, very slow and very high cost compared to simultaneous ICP instruments.

A simultaneous instrument (Figure 12) splits light in two dimensions measuring all wavelengths at once on a CCD or CID chip. These instruments enable measurement at any wavelength enabling measurement at both high and low concentrations. They are very high speed. All elements amenable to ICP (up to 72) can be analyzed in 1 – 2 minutes. Simultaneous instruments are the best for environmental testing because of their lower cost and higher throughput. However, they do suffer from inter-element interferences requiring high pixel chips and some mathematical treatment of data to compensate for interferences.

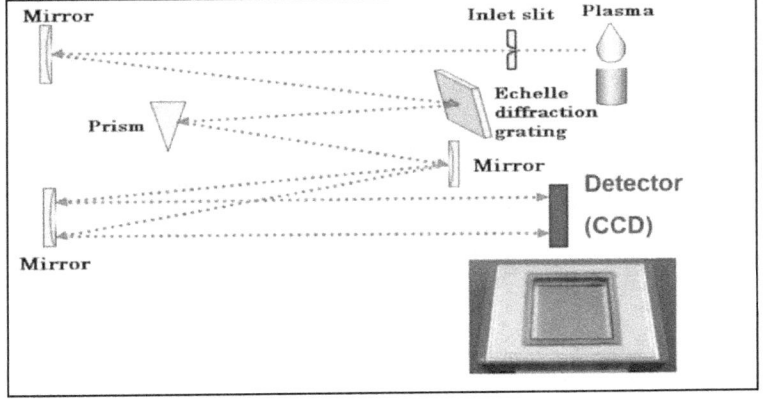

Figure 12, Simultaneous ICP-AES with Shimadzu CCD detector shown

The spectra on the right in Figure 13 represents full separation of one wavelength from the other. The spectra on the left does not. It is possible, in the spectra on the left for the shoulder of a very large peak to interfere with the other wavelength.

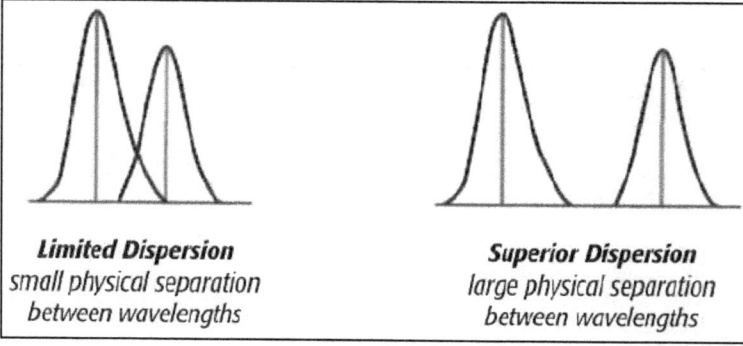

Figure 13, Limited Dispersion and Superior Dispersion, Shimadzu

Each ICP has a source, or plasma, optics to split the light into its various wavelengths, and a detector to measure each specific wavelength and its intensity. The position of the light on the detector determines its wavelength and the intensity is proportional to concentration. To get the sample into the plasma it must be nebulized (Figure 14). Nebulization means creating very small droplets that can be carried by the argon carrier gas into the plasma. To do this, the sample is aspirated, or pumped into a nebulizer. It enters a spray chamber that removes the largest droplets. Only a fine mist containing about 1% of the sample aspirated makes it into the plasma. 99% of the sample aspirated goes down the drain.

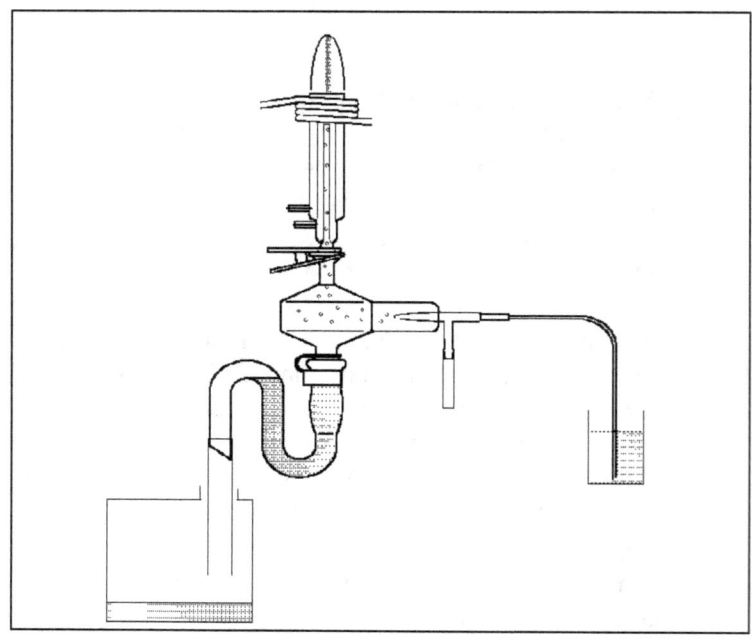

Figure 14, Shimadzu ICPE-9800 Nebulizer and Torch assembly

You can increase the amount of sample to the plasma; however, there are always tradeoffs. High dissolved salts in the sample can clog the nebulizer, and with every sample there are always elements you may not be interested in that can interfere. Plasma is gaseous argon at about 10,000 C. Elements that are carried into the plasma ionize. In ICP emission, the photons emitted during ionization are measured. The wavelength (or energy) of the photon is specific to the element ionized. The number of photons emitted is proportional to the number of atoms that make it to the plasma (concentration).

The plasma torch (Figure 14) consists of three tubes, usually made of quartz. There is an outer tube, a middle tube, and the sample injector. Argon gas passes between the outer tube and the middle tube. Seen here as swirling around in a spiral flow. A second stream of argon, at a much slower flow rate, passes between the middle tube and the sample injector. This second gas stream, also called auxiliary gas, changes the position of the plasma relative to the injector. A third gas flow, the nebulizer gas, carries the sample through the injector and into the plasma. This stream makes a tunnel through the center of the plasma. As a note, although argon is usually used, other gases can be used for specific purposes. A load coil, made of copper, surrounds the top of the torch and is connected to a Radio Frequency generator. When power is applied (usually about 1100 W), an alternating current oscillates at the frequency of the generator (27 or 40 MHz). The oscillation of current creates a magnetic field in the area inside the coil at the top of the torch. With argon flowing through, a high voltage spark is applied ionizing some of the argon. This causes a chain reaction within the magnetic field breaking down the argon gas to contain argon gas, argon ions, and electrons. This is an inductively coupled plasma. The viewing area for ICP emission work is somewhere between the 6500C and the 6000C. The argon plasma for ICP-MS is slightly different; however, for our purposes this discussion should be sufficient.

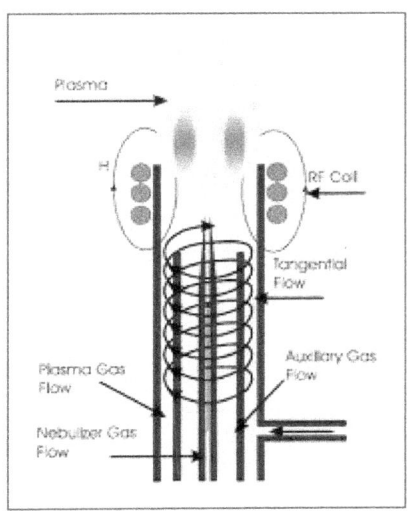

Figure 14, Schematic of ICP torch[11]

---

[11] https://www.uspto.gov/web/patents/classification/cpc/image/cpc-definition-H05H/media2.jpg, accessed February 24, 2017

To summarize, samples in a liquid (usually aqueous) state are introduced into a plasma. Elements in the sample ionize and release radiation. The wavelength of the radiation is specific to the element. The intensity of the radiation is proportional to the concentration.

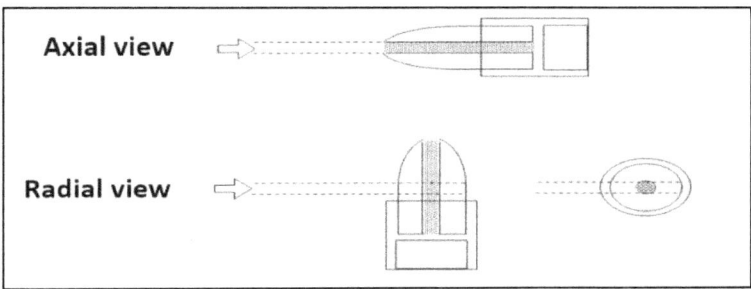

Figure 15, axial and radial view

Traditional ICP, or method 200.7, uses a radial view. The newer method, Method 200.5, is an axial view method. The 2012 method update allows both axial and radial viewing (Figure 15 and Table 11) for EPA 200.7 in wastewater. Because of its longer viewing path, axial view has better detection limits. It also suffers from more interferences and is not as linear. Radial view is less affected by interferences with a large dynamic range. If your samples will always be of concentrations high enough, then get an instrument with only a radial view. Otherwise, get one with dual view.

Table 11

Method	Analyte	View
EPA 200.7/6010	Most metals	Radial, with axial allowed at Part 136.6
EPA 200.5	Trace metals	Axial

A modern simultaneous ICP-AES splits light emitted by the plasma first into one dimension, perhaps horizontally, at a grating (or prism) and then splits again, in this case vertically, creating a two-dimensional pattern designed to cover as much as the detector surface as possible. The position of light hitting the detector determines the wavelength and the intensity is proportional to concentration.

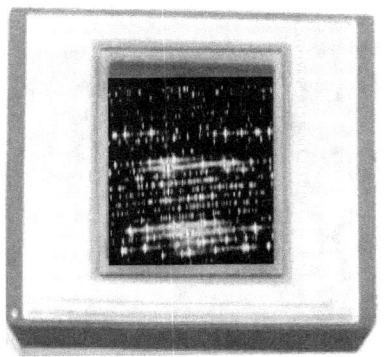

Figure 16 Charge Coupling Device (CCD)

Figure 16 is an example of a Charge Coupling Device (CCD) chip with emission spectra illuminated on it. The idea is to have a chip large enough so that the entire spectra is recorded. You also want as many pixels as possible for highest resolution. Better resolution provides better separation of wavelengths.

Figure 17, Segmented CCD or SCD

Figure 17 is a picture of a segmented CCD, or SCD, composed of 70 small CCD's. These measure only in regions where there is a chip and may be slower in samples containing both high and low concentrations of elements.

Figure 18, Charge Injection Device (CID) chip

A CID detector (Figure 18) is smaller with a phosphor coating that enables measurement at the lower wavelengths. CID's can be 10 – 100 times noisier than CCDs. A report by Test America mentioned false readings for lead caused by silicon and copper. These interferences need to be corrected by inter-element correction software[12].

---

[12] https://doecap.oro.doe.gov/EDS/Readings/207%20Larry%20Penfold-Optical%20ICP%202011.ppt

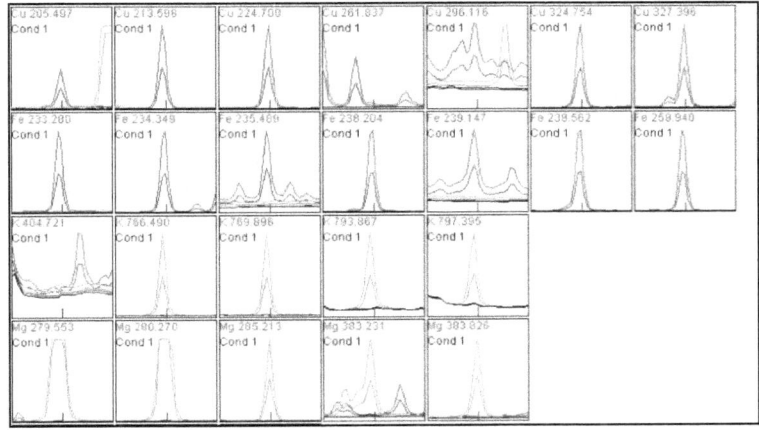

Figure 19, Spectra collected on a modern ICP-AES, Shimadzu

The software processes the spectra as individual peaks (Figure 19). Keep in mind that these are emission spectra. You do not integrate peaks. You measure peak height at the maximum.

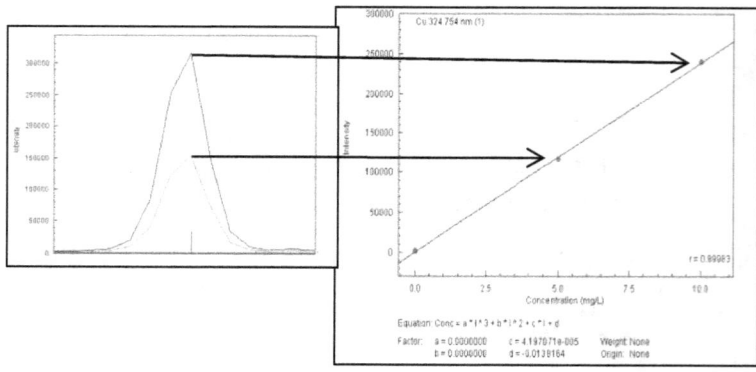

Figure 20, calibration from ICP emission spectra

Figure 20 is a calibration curve created by measuring the maximum intensity for each standard at the peak maxima.

A horizontal torch orientation clogs easier and increases memory effects requiring longer washout times (Figure 21).

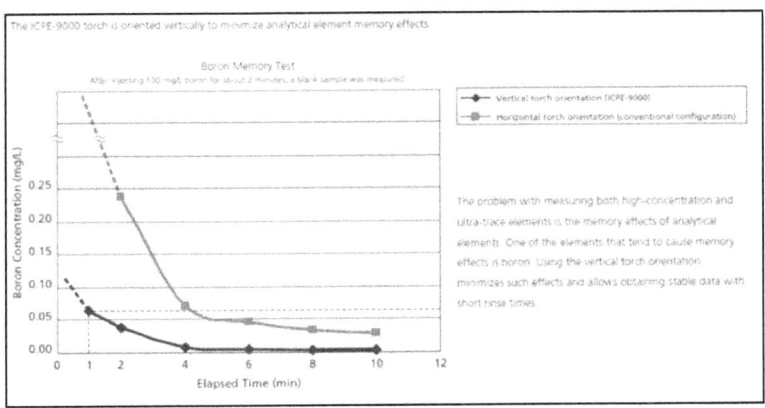

Figure 21, Washout test on the Shimadzu ICPE-9000

To measure at the lower wavelengths, you cannot pass spectra through air. You must either purge the optics chamber with argon gas or empty it using a vacuum. Figure 22 shows an advantage of vacuum purge as it relates to start up time.

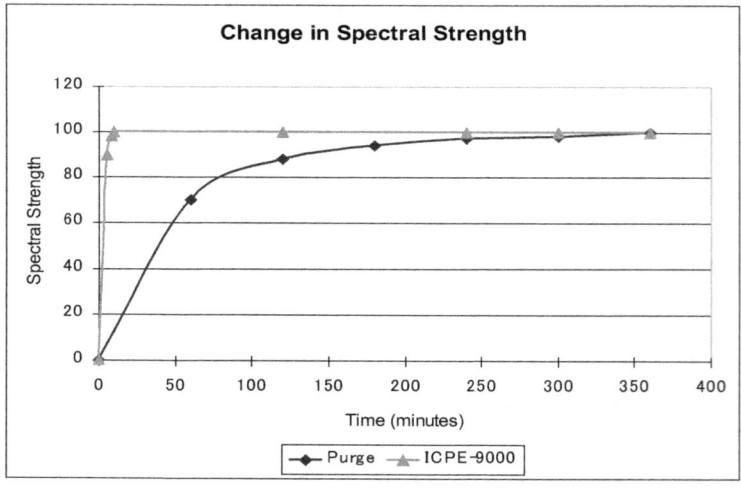

Figure 22, Time to get a stable signal

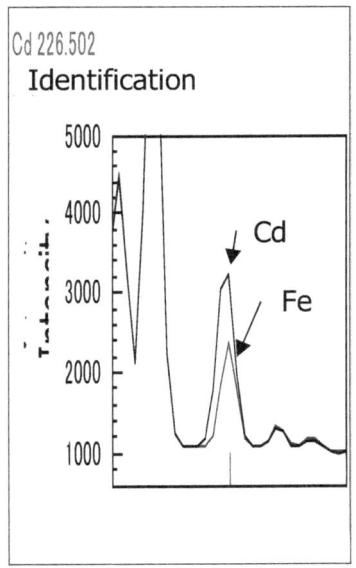 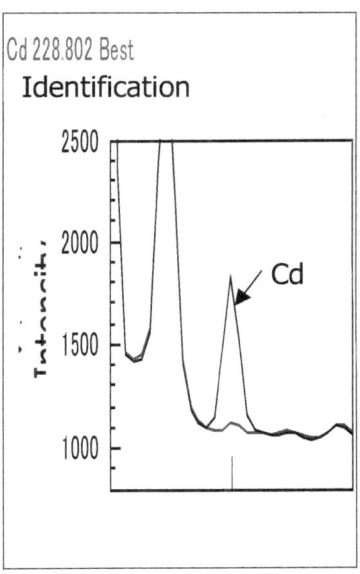

Figure 23 Iron interference on Cadmium

Figure 23 is an example showing the spectral interference of iron on cadmium; if iron is in the sample, you cannot trust a cadmium result at 226.502 nm. However, software can warn you that Fe interferes and tell you that the 226.502 result is likely incorrect. It can suggest you only use the 228.802 result. Alternatively, it can perform an Inter-element correction based on an analysis of pure iron standards. These corrections can be made after the sample was run.

All manufacturers provide various torch configurations (Figure 24) depending on what you are trying to do.

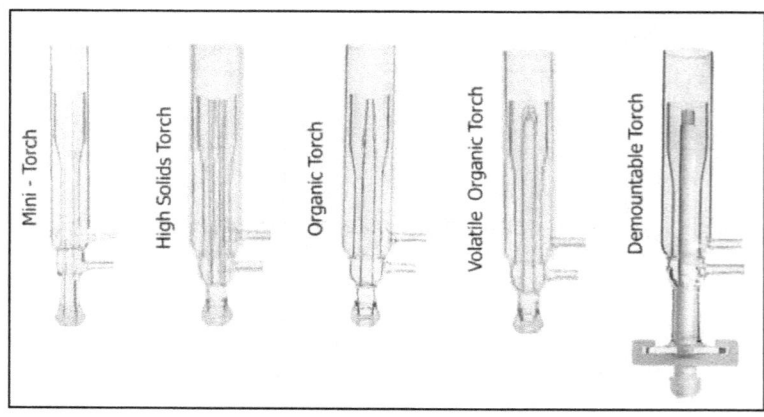

Figure 24, various torches

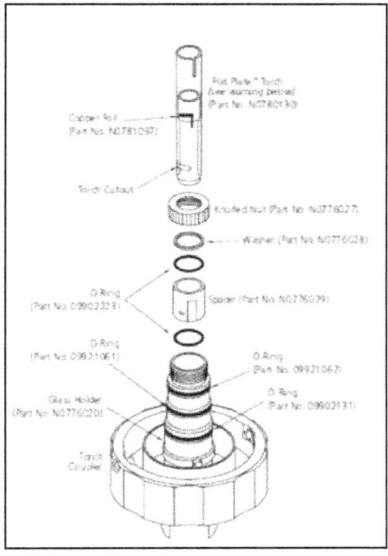

Figure 25, cassette torch[13]

---

[13] https://www.perkinelmer.com/lab-solutions/resources/docs/Optima%207300%20V%20Consumables%20Reference%20Guide.pdf, accessed February 24, 2017

Some manufacturers have cassette torches (Figure 25) designed to minimize labor to replace a torch. High TDS samples can be run by wetting the argon (Figure 26) just prior to introduction of the sample,

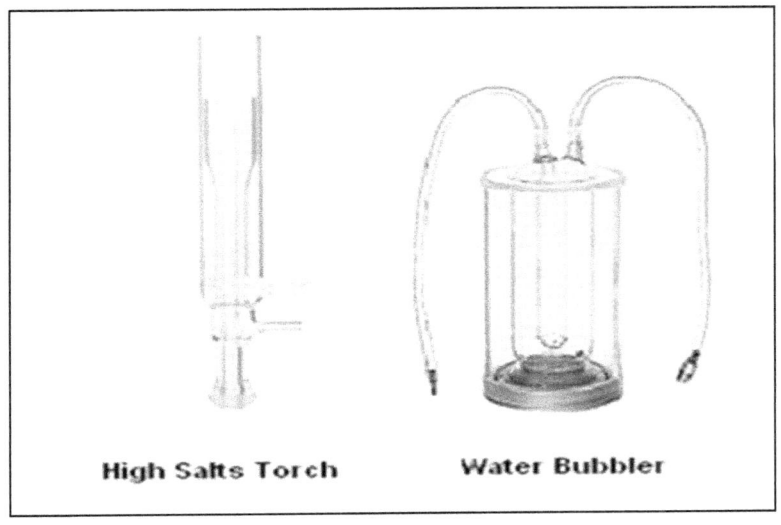

Figure 26, High salts kit (Shimadzu)

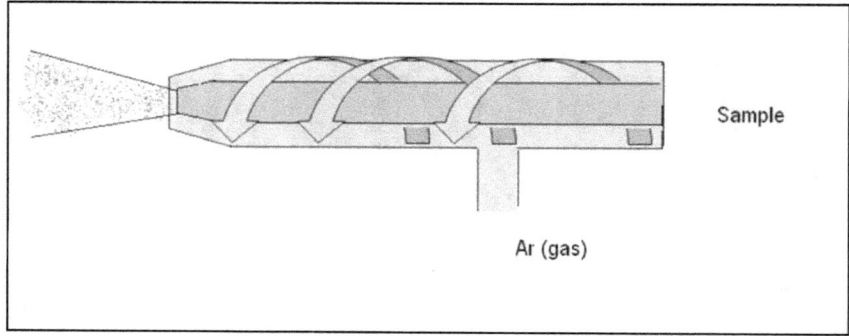

Figure 27, conical nebulizer

Conical Nebulizers (Figure 27) are 1 – 2 % Efficient.

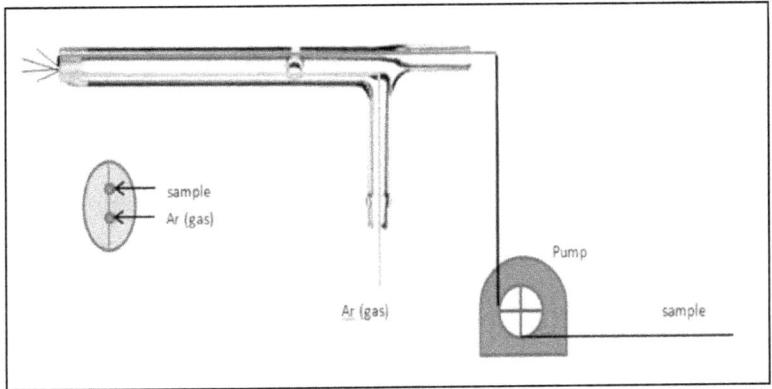

Figure 28, V-Groove Nebulizer

V-Groove Nebulizers (Figure 28) are better for high solids samples, but not as efficient as conical nebulizers are.

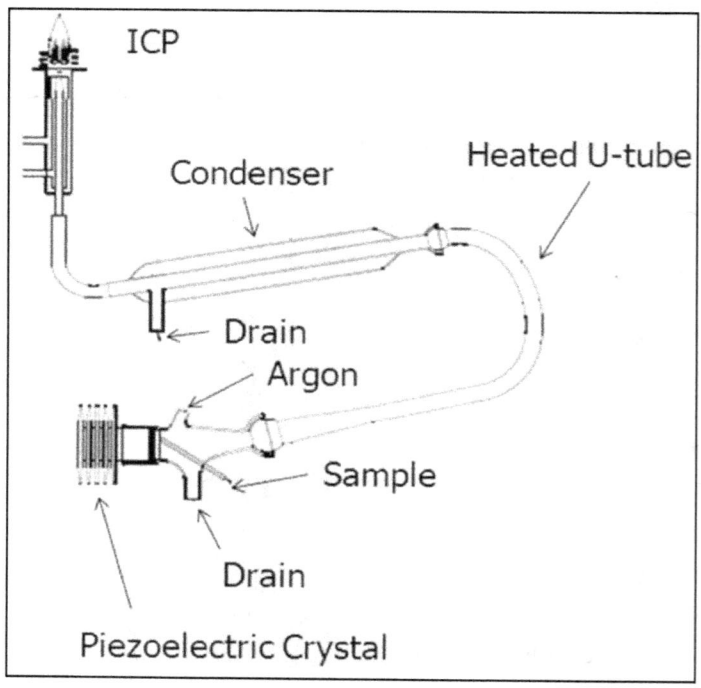

Figure 29, Ultrasonic nebulizer increases efficiency and lowers detection limits.

With an ultrasonic nebulizer (Figure 29), sample is aspirated using a peristaltic pump. The sample hits a piezoelectric crystal vibrating at a million vibrations per second. The vibrations convert about 10% of the sample into an aerosol. The aerosol is heated to prevent loss. Then cooled just before introduction to plasma to remove moisture. Then the aerosol enters the plasma. The 10% increase in nebulization efficiency decreases detection limits by a factor of 10. This makes analysis of low concentrations of Pb, As, Se, and Hg possible by emission ICP.

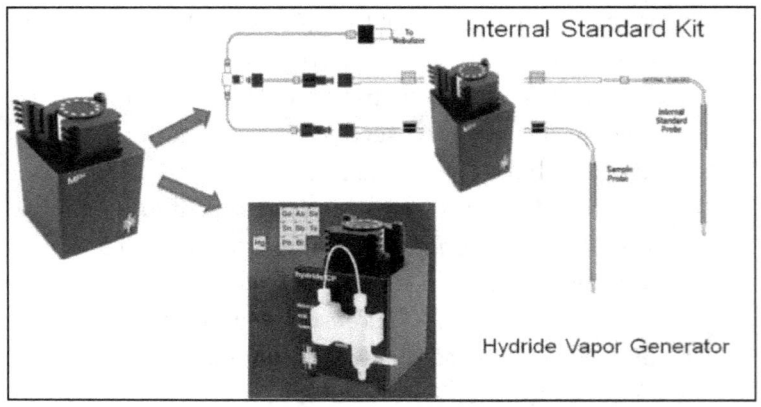

Figure 30, Shimadzu Hydride generation kit

Hydride generation (Figure 30) measures very low concentrations of As, Se, Hg, by ICP-AES

Suggestions on selecting ICP-AES instruments

Tabulate these things as you look for an ICP. Resolution is in terms of pm. The smaller the pm the higher the resolution:

1. High pixel CCD
2. High resolution
3. Low argon consumption
4. Short start up time
5. Vertical torch
6. Axial and radial view
7. Long-term stability
8. High sample throughput
9. Good technical support
10. Training on environmental methods
11. Prompt service

Inductively Coupled Plasma – Mass Spectrometry

ICP- mass spectrometry (Figure 31) is used for measuring metals in environmental samples by EPA Method 200.8 and 6020. It can replace graphite furnace and hydride generation AA. It drastically improves your labs throughput because multiple elements can be analyzed in a single aspiration. As of now, all ICP-MS instruments have a horizontally oriented torch. They are limited in TDS to about 2000 mg/L. Some manufacturers claim a higher TDS tolerance, but some sort of dilution does this.

Figure 31, Schematic of an ICP-MS[14]

14

http://biochem.pepperdine.edu/dokuwiki/lib/exe/fetch.php?media=chem331:icp-ms.jpg, Accessed February 24, 2017

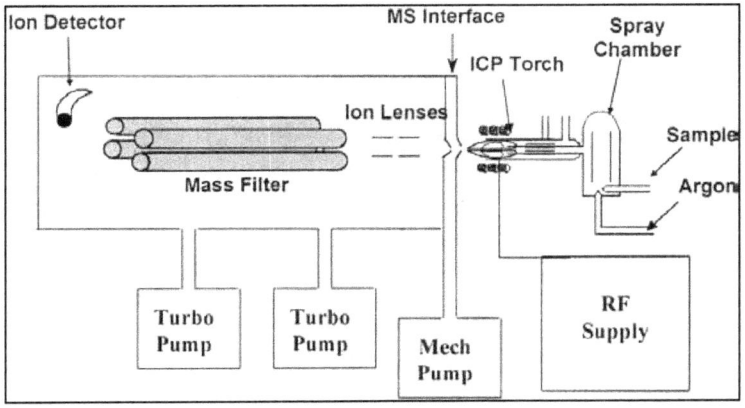

Figure 31, Schematic of an ICP-MS[15]

15

http://biochem.pepperdine.edu/dokuwiki/lib/exe/fetch.php?media=chem331:icp-ms.jpg, Accessed February 24, 2017

Sample introduction is very similar to that previously described for ICP-AES. Elements are ionized and turned into positive ions. The ions exit the plasma at the interface and pass through a lens into a vacuum. The vacuum and lenses focus the ions into a mass-selective device (quadrupole) that separates the ions based on their mass/charge ratio. A data system interprets the signals as ions hit the detector. ICP-MS can provide qualitative results, but the instrument is usually calibrated to provide both qualitative and quantitative analysis.
Environmental methods target certain analytes.

Spectral interferences are caused by polyatomic ions, doubly charged ions, or other ions that have the same mass as the analyte. The effect is that a spectral interference always results as an increase in signal. This increase is independent of the target ions concentration. Physical interferences, caused by high viscosity or differences in salt content between the sample and the standards are common in ICP-AES and ICP-MS. The spectral interferences here are different from those described for ICP-AES. Here we are concerned with other elements or combination of elements that have the same mass (not wavelength) as the element of interest. Physical and ionization interferences may be corrected by the internal standard technique. Thus, internal standards are used with ICP-MS. Spectral interferences can be eliminated by selection of a non –interfering mass, or by use of collision or reaction cells.

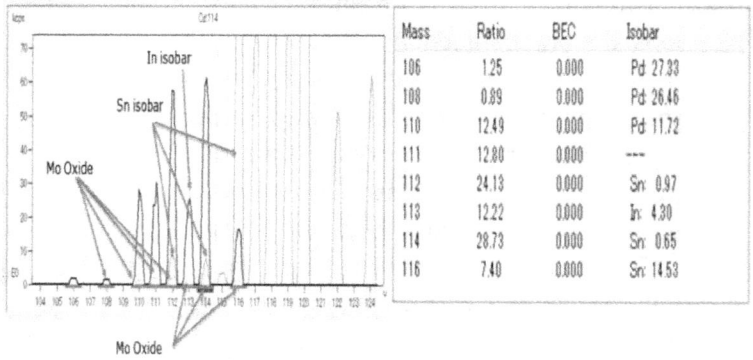

Figure 32 is an example of Cd analysis.

Cadmium has eight isotopes, or potential masses. However, tin, palladium, and Indium all have masses equal to at least one cadmium mass. Molybdenum oxide (a polyatomic) has a mass of 111. If any of these elements are in the sample, they will be measured as a positive interference on cadmium.

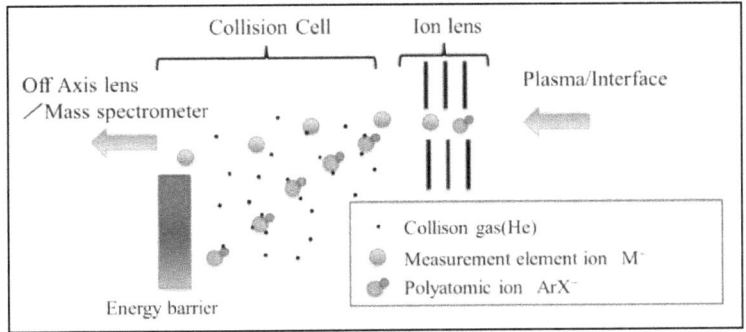

Figure 33, Example of a collision cell

A collision cell (Figure 33) can reduce polyatomic interferences, such as the molybdenum oxide interference on cadmium. The collision cell bombards ions with helium gas preventing the polyatomic ions from passing into the detector.

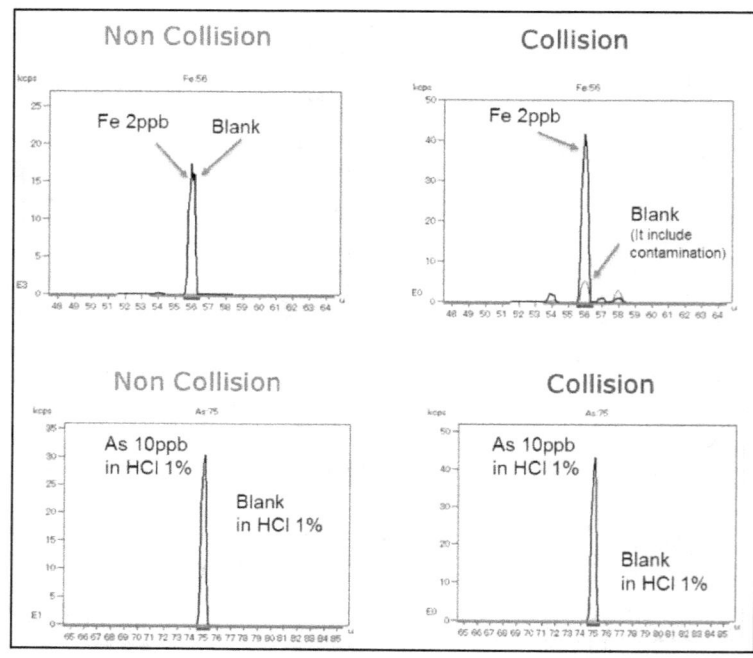

Figure 34, examples of spectra before and after a collision cell

Figure 34 are examples of how a collision cell removes the positive interference of polyatomic ions.

Table 12 Interfering masses

Interferent	Mass	Element
ArH^+	38	K
Ar^+	40	Ca
ArC^+	62	Cr
ArO^+	56	Fe
Ar_2^+	80	Se

Table 12 lists examples of molecules that can cause positive interferences if there is no collision cell.

Without collision cells your optional solution to pollution is dilution. Alternatively, to not report elements that may have an interference and run them by graphite furnace or hydride AA. For example, if chloride interference prevents you from measuring Arsenic, then you will have to measure Arsenic some other way.

Unfortunately, the collision cell is not approved for drinking water reporting. This requires that each sample be examined for potential spectral interferences and corrected with equations provided in Method 200.8. Fortunately, most manufacturers will do this automatically with their software (Figure 35).

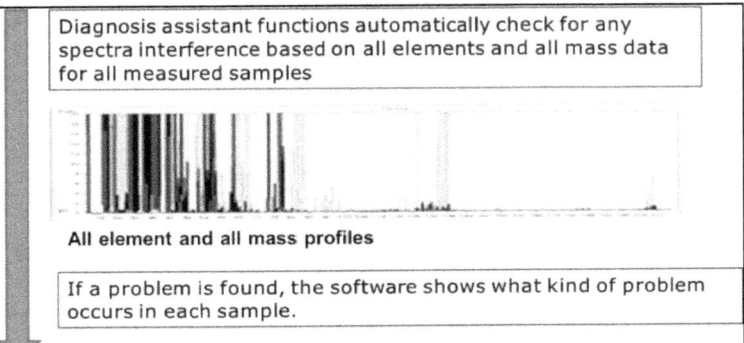

Figure 35, Shimadzu software showing diagnosis for interferences by ICP-MS

Suggestions on selecting ICP-MS instruments

Precision on the ICP-MS is usually 1 – 3% and increases to 3 – 5% as the peristaltic pump tubes flatten or as matrix begins to build up on the cones. The higher the TDS of your samples the faster precision will degrade. These are the things that cause precision to get worse:

- Pulsations and fluctuations of peristaltic pump
- Blockage of the nebulizer
- Poor drainage
- Buildup of solids on injector
- Erosion of orifice from acids
- Blockage of cone orifice
- Coating of ion optics from matrices

Watch out when manufacturers talk about capability of high TDS and then show sodium chloride solutions. These are great for proving a collision cell works for Arsenic, but do not show much for longevity with high TDS. It is the calcium, magnesium, iron, aluminum, etcetera that causes the problems. Look for data in real samples, or in synthetic seawater.

Think of these things when compiling your list of needs for an ICP-MS. The good thing is, modern ICP-MS are so sensitive you will be able to meet MDL requirements of Method 200.8. Remember that you cannot use a collision cell for drinking water, but you can with wastewater.

1. What is the TDS of your samples?
2. What detection limits do you need?
3. Do you need a collision cell?
4. Can you use a collision cell?
5. Is the software easy to use?

You can ask for proof that the instrument will run 200.8 but you can feel confident that it can. See what applications there are for complicated samples that will require dilution. How well does it do in digested samples? Check the MDL data. Often manufacturers only provide IDL, or only report data run on "clean" samples. How hard is it to run the thing? Will the vendor provide a method, or will you have to figure out the software, make a method, and validate it before you can start running samples? How steep is the software learning curve?

Chromatography

Chromatography uses a column to separate compounds from each other and measure them. Gas chromatography separates analytes when they are in the gaseous phase, ion chromatography separates ions in water, and liquid chromatography separates organics compounds either in aqueous phase or from an organic phase. For gas chromatography, most environmental samples are first extracted into an organic solvent and then concentrated to a small volume. Ion chromatography usually injects the sample directly, HPLC may inject directly or after an extraction.

Gas chromatography is quite simple, consisting of an injector, the separation column, and a detector. The column separates the components of a mixture and the identity of each component can be made by the time it is measured by the detector.

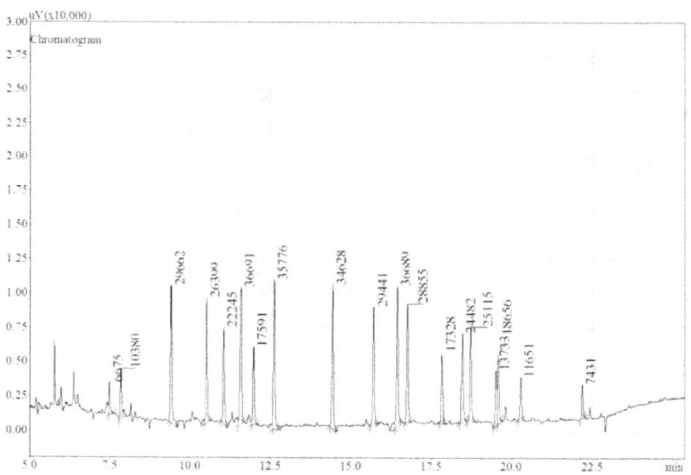

Figure 36, Example chromatogram

Figure 36 is a chromatogram showing the elution of a number of peaks. The time that a peak comes out identifies the compound and the intensity of the peak is proportional to its concentration. All chromatography methods rely on separation by time and then detection of the peaks. Different types of detectors can be used depending on what you are trying to measure. Since chromatographic measurements identify peaks only by time, it is useful in very complex samples to use different types of detectors on the same sample. The different detectors help to confirm the identity of the analyte. In addition, specific detectors, respond only to certain kinds of analytes.

Gas Chromatography (GC)

Before you buy a GC, you must know which method you are going to run. The method determines the detector. Since GC methods require two-column confirmation, you need to decide how many identical detectors you need, for a small sample load; you can get one detector and move the column. For less hassle and larger sample loads, get two detectors, two columns, and two injectors.

Things to consider before buying a GC:

1) What are the analytes/methods you will run?
   a. This determines the detectors
2) How many detectors do you need?
3) What kind of throughput do you need?
4) How does the GC control flow?
5) Are spare parts and service readily available?
6) Will the instrument be discontinued soon?
7) What is the software like?

Table 13 are typical environmental GC methods.

Table 13 Environmental GC methods

Method	Analyte	Detector
EPA 608	Organochlorine pesticides	Electron capture
8020	BTEX	Purge and Trap with PID

Gas Chromatography Mass Spectrometry (GCMS)

Gas Chromatography mass spectrometry is the workhorse of the organic environmental laboratory. This is the instrument you will use for methods such as 524.2, 624, 8260 volatiles, and 525, 625, and 8270 semi-volatiles. GCMS has a long history in the environmental laboratory and are often considered a commodity. There have been advancements in detection technology that are held back, to some degree, by regulation. A basic GCMS consists of a GC with an injector, and column. The detector is a mass selective device, usually a quadrupole. There are subtle differences in the mass spectrometry detectors of each manufacturer. Each instrument is capable of running the EPA methods. There are differences in how the quadrupoles are manufactured, how the ionization is done, and detector design. None of these differences take away from basic functionality; the ability to run the methods. Other differences include access to regularly maintained parts and of course, intangibles, such as service and technical support.

The mass spec detector (Figure 37) ionizes the sample. In most environmental methods, the ionization is by electron impact at 70 electron volts. The ionization splits the analyte into a distinctive, very reproducible mass fragments. These fragments are accelerated through a scanning mass-selective device to a detector. The data output is a fragmentation pattern that can be library searched. Each compound has its own pattern of mass and relative abundance. Quantitation is made by measuring intensity of mass. Qualitative ID is made by retention time and how well the fragmentation pattern matches the pattern of a known standard or library.

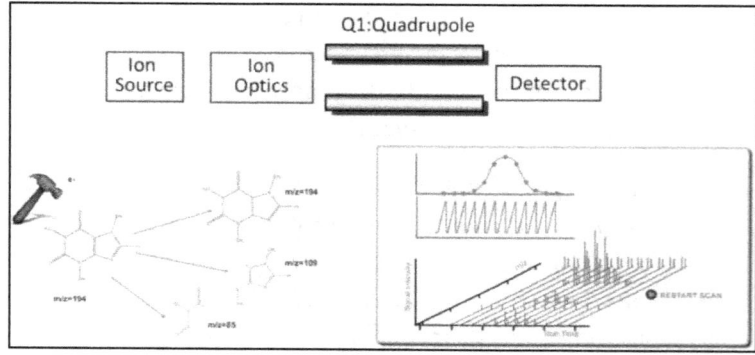

Figure 37, schematic of ionization and single quadrupole mass spectrometer

Before you can run environmental methods by GCMS, you must adjust the mass fragment pattern so that all laboratories obtain reproducible spectra. You adjust the "tune" to meet method criteria for BFB (volatiles) or DFTPP (semi-volatiles). Older instruments did not do so well at higher mass, so newer instruments may have difficulty matching the correct tune criteria, especially at higher masses. The fact that the older methods require an "incorrect" mass-spectral pattern for BFB and DFTPP is irrelevant. You must "tune" your instrument to meet these method criteria. Make sure, from the manufacturer, that the instrument can meet the EPA tune criteria for the method you need to run.

The BFB criteria for EPA volatiles are similar with a few notable exceptions. Table 14 compares the BFB relative abundance criteria for six different EPA methods. Method 524.3/524.4 is the least stringent and methods 524.2, 624, and 8260c are the most stringent. If you always pass the most stringent criteria, you will also pass the least stringent.

Table 14 BFB Tune criteria for different EPA methods.

Mass (m/z)	Relative Abundance Criteria				
	Method 524.2	Method 524.3 /M524.4	Method 624	Method 8260C	CLP-SOW
50	15 to 40% of 95	NA	15 to 40% of 95	15 to 40% of 95	15 to 40% of 95
75	30 to 80% of 95	NA	30 to 60% of 95	30 to 60% of 95	30 to 80% of 95
95	Base Peak, 100%	Base Peak, 100%	Base Peak, 100%	Base Peak, 100%	Base Peak, 100%
96	5 to 9% of 95	5 to 9% of 95	5 to 9% of 95	5 to 9% of 95	5 to 9% of 95

173	<2% of 174	<2% of 174	<2% of 174	<2% of 174	<2% of 174
174	>50% of 95	>50% of 95	>50% of 95	>50% of 95	50 to 120% of 95
175	5 to 9% of 174	5 to 9% of 174	5 to 9% of 174	5 to 9% of 174	4 to 9% of 174
176	>95 to <101% of 174	>95 to < 105% of 174	>95 to <101% of 174	>95 to <101% of 174	95 to 101% of 174
177	5 to 9% of 176	5 to 10% of 176	5 to 9% of 176	5 to 9% of 176	5 to 9% of 176

Table 15 compares the DFTPP for five different EPA methods. Method 525.3 is the least stringent followed by Method 525.2. Methods 625 and 8270D are the most stringent. Similar to BFB, if you pass the criteria for the most stringent method, you will also pass for the least stringent methods.

Table 15, DFTPP tune criteria for different EPA methods

Relative Abundance Criteria

Mass (m/z)	Method 525.2	Method 525.3	Method 625	Method 8270D	CLP-SOW
51	10 to 80% of 198		30 to 60% of 198	30 to 60% of base peak	10 to 80% of 198
68	<2% of 69	<2% of 69	<2% of 69	<2% of 69	<2% of 69
69		present			
70	<2% of 69	<2% of 69	<2% of 69	<2% of 69	<2% of 69
127	10 to 80% of 198		40 to 60% of 198	40 to 60% of 198	10 to 80% of 198
197	<2% of 198	<2% of 198	<1% of 198	<1% of 198	<2% of 198
198	Base peak, or	Present*	Base peak, 100 %	Base peak, 100%	Base peak, 100 %

	>50 % of 442		relative abundance	relative abundance	relative abundance
199	5 to 9% of 198	5 to 9% of 198	5 to 9% of 198	5 to 9% of 198	5 to 9% of 198
275	10 to 60% of base peak	NA	10 to 30%of 198	10 to 30% of 198	10 to 60%of 198
365	>1% of 198	>1% of Base Peak	>1% of 198	>1% of 198	>1% of 198
441	Present, but <443	<150% of 443	Present, but <443	Present but < 443	Present, but <443
442	Base peak or >50% of m/z 198	Present*	> 40% of m/z 198	>40% of 198	>50% of m/z 198
443	15-24% of mass 442	of mass 442	17-23% of mass 442	17-23% of mass 442	15-24% of mass 442

Suggestions on selecting a GCMS
1) Will it easily tune to DFTPP or BFB?
    o How long will it stay tuned?
2) How often do you need to recalibrate?
3) How often must you clean the source?
4) Is the source easy to clean?
5) How hard is it to clean the source?
6) How fast does the MS scan?
7) Are spare parts, technical support, and service readily available?
8) Will it be discontinued soon?
9) Is the software easy to use?
    o How different is it?
    o Can the quality control functions be automated?
    o Will it link to a LIMS?

Stability is very important. The EPA methods require that internal standards do not drift significantly during the analytical batch. Get stability/drift (Figure 38) data from the manufacturer if possible.

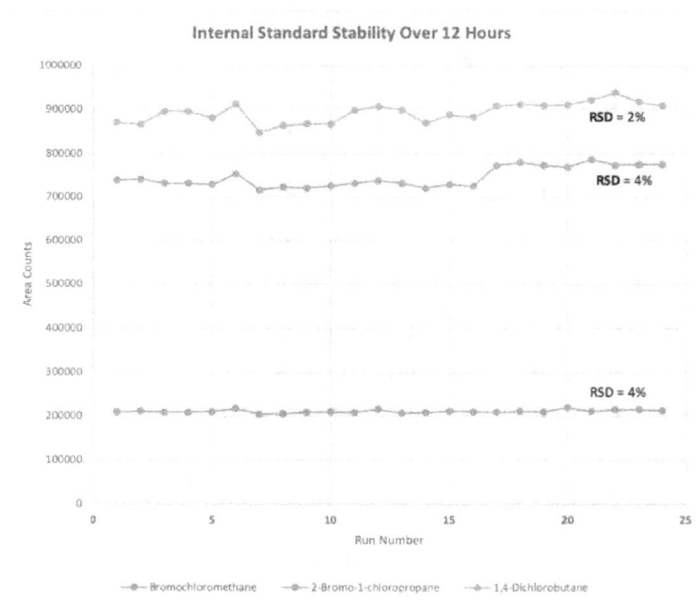

Figure 38, Internal Standard Stability

Scan speed is important to enable you to integrate as many points across a peak as possible. The fewer number of points per peak, the worse the precision. EPA methods were written for packed columns. Many say 5 scans per peak. We recommend 15 – 20 scans per peak. Newer methods are allowing selected ion monitoring (SIM). SIM provides greater sensitivity by dwelling at one mass longer than in full scan. A fast scanning instrument is capable of full scan and SIM in the same method at the same time. High speed scanning enables you to use shorter columns with a narrower bore to shorten run times. The smaller column results in much narrower peaks, to still integrate 10- 15 points per peak you need to scan faster. Figure 39 is an 8270 run in 12 minutes.

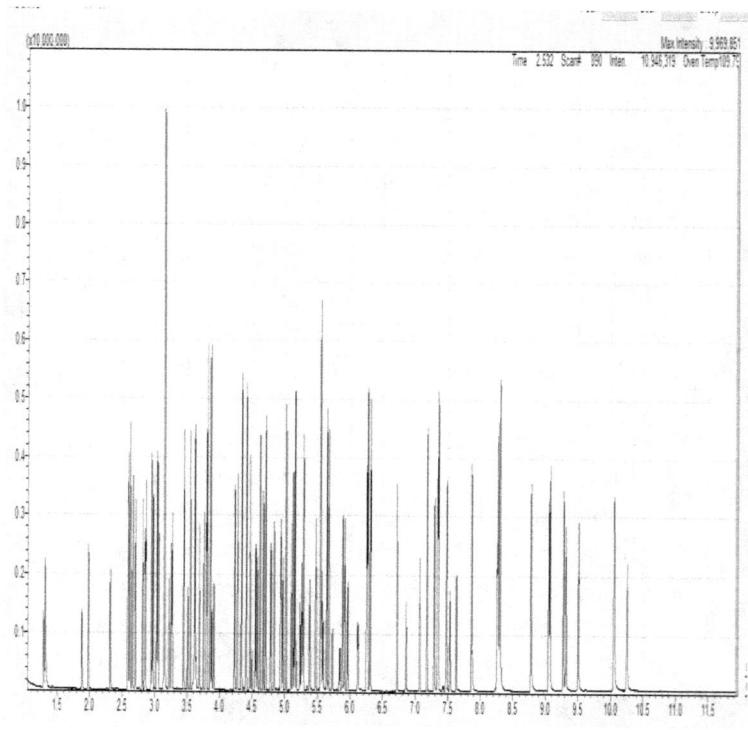

Figure 39, Rapid 8270 run

Features to compare on GCMS Systems:
1. Are the quadrupoles all-metal or are they metal strips mounted on ceramic?
2. Do the quadrupoles have easily cleaned pre-filters that will get dirty before the quadrupoles do?
3. Is the source easily cleaned, deactivated, or inert?
4. How many filaments does the source have?
5. Is the source accessible?
6. What type of detector is it?
7. How does the detector minimize noise?
8. What kind of vacuum pump does it have and what is the pump-down rate?
9. What is the maximum scan speed? This is most important for fast analysis of semi-volatiles.
10. Does it have the capability to do chemical ionization?

Critical factors to compare:

1. Technical support
2. Service
3. Durability of hardware
4. Quality of manufacture
5. Reputation of manufacturer

Table 16 Gas Chromatography Mass Spectrometry Methods

Method	Analyte	Sample Preparation
EPA 624/8260	Volatiles	Purge and Trap
EPA 625/8270	Semi-volatiles	$MeCL_2$ extraction and concentration

Triple Quadrupole GCMS

Triple quadrupole GC-MS/MS a new technology for targeted environmental analyses, due to high level of selectivity and sensitivity. A triple quadrupole operates very similar to single quadrupole except that a collision cell and another quadrupole is added. Technically then, a triple quadrupole is only a double quadrupole. The extra quadrupole after the collision cell selectively eliminates almost all noise. Notice that in Figure 40, the signal from the triple quadrupole (MRM) is lower than the SIM signal. Note though, that after the collision cell and second quadrupole there is almost no noise. This elimination of noise significantly increases the signal to noise ratio lowering the detection limit of MRM relative to SIM.

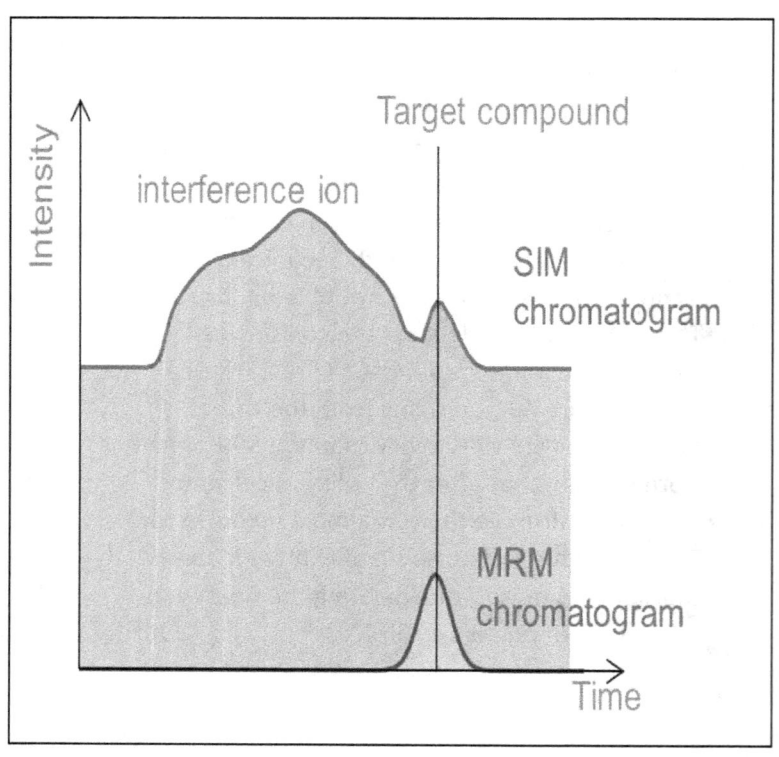

Figure 40, illustration demonstrating noise reduction by MS/MS

For an MRM transition, the compounds eluting from the column are first fragmented by the source, just as in regular single quadrupole GCMS. One of the fragments passes through the first quadrupole as a SIM ion. This SIM ion is a precursor ion. The precursor ion passes through the collision cell and is bombarded with collision gas (usually helium) at a predetermined energy to fragment it into product ions. The product ions pass through the second quadrupole as SIM ions. One product ion is called SRM and multiple product ions measured are called MRM. Using multiple ions enables one ion to be the quant ion and other ions to be qualifier ions.

Figure 41, precursor and product ions

In Figure 41, the precursor ion is the molecular mass [M$^+$]. Two chlorine atoms are lost in the collision cell. The product ion is [M-2Cl]$^+$.

Figure 42 shows the tremendous selectivity and noise reduction of MRM compared to SIM. The chromatogram on the left is a SIM ion at mass 290. The chromatogram on the right are two MRM product ions after collision of the 290 ion. Note the absence of baseline, and extremely higher S/N ratio.

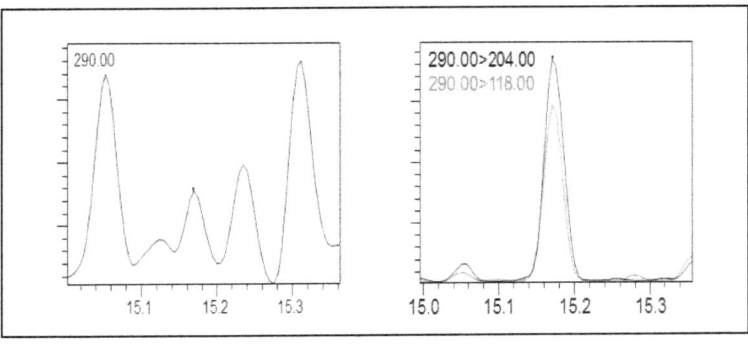

Figure 42, comparison of SIM and MRM

Figure 43 is an example Dieldrin is fragmented in the source at 70 electron volts

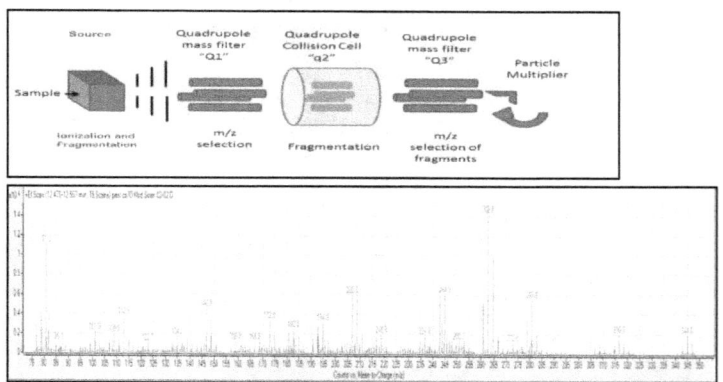

Figure 43 fragmentation of Dieldrin

In Figure 44, a precursor chosen for dieldrin passes through the first quadrupole in SIM mode

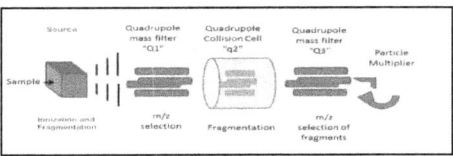

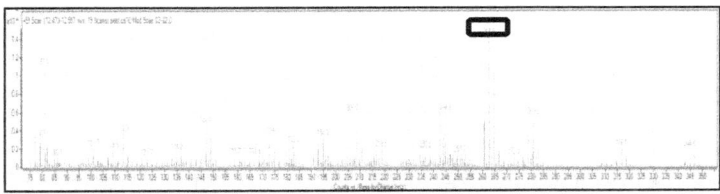

Figure 44, a precursor of Dieldrin passes to collision cell

Collision energies optimized for each target precursor generate the greatest abundance of a desired product m/z(s). Precursor ions contributed from different compounds (target vs matrix) will have a unique fragmentation pattern based on their structures.

The second quadrupole selectively passes the product ions that were determined by the target structure. Therefore, the resultant product ion spectrum (Figure 45) is almost entirely due to the target precursor ion and not the chemical background.

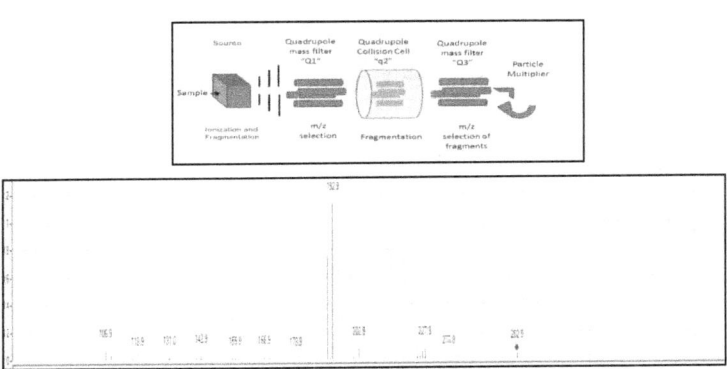

Figure 45, Product ion spectra of Dieldrin

Figure 46 is a product ion chromatogram of method 625 and 608 compounds. Each different color represents a different mass.

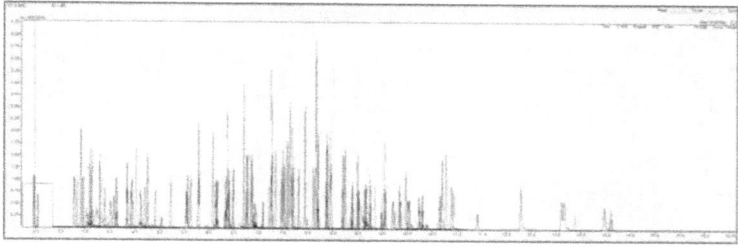

Figure 46, product ion MRM chromatogram of Method 625 semi-volatiles and pesticides

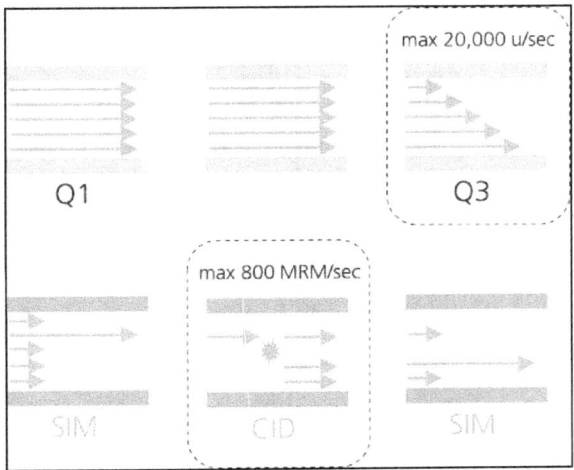

Figure 47, operating the triple quadrupole in single quadrupole mode

An advantage of some triple quadrupole instruments is that you can also use them to run your semi-volatiles in full scan mode (Figure 47). You can essentially turn off the first quad and the collision cell.

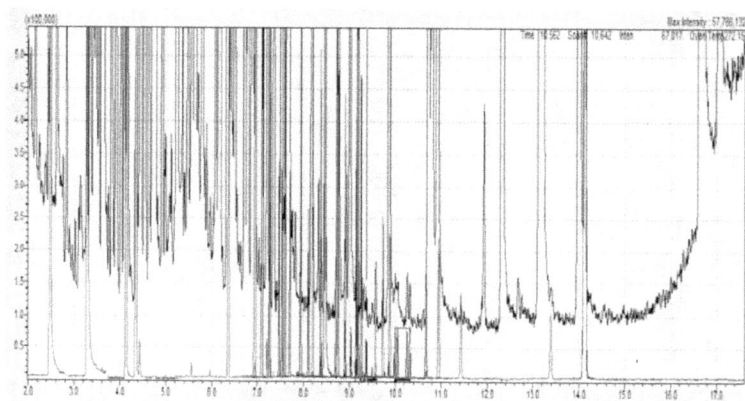

Figure 48, Full scan and MRM chromatogram

Figure 48 is an example of the sensitivity of the triple quadrupole and how well it eliminates the matrix. This shows 5 ppb pesticides by MRM buried under a 2-ppm chromatogram of method 625 semi-volatiles.

Ion Chromatography

Ion chromatography is a form of liquid chromatography used to separate ions in aqueous solutions. Typically, IC analyzes anions such as fluoride, chloride, nitrite, bromide, nitrate, phosphate and sulfate. Cations by IC was approved for wastewater reporting in the 2012 method update. Ion chromatography can detect other ions as well. Chromate, perchlorate, and organic acids are examples. Because different ions can be detected, good separation is necessary to avoid overlapping peaks of analyte and contaminants. Detection using conductivity is common; however, UV can detect some anions, such as nitrite and nitrate. UV is useful in solutions containing high concentrations of chloride. In these solutions the chloride peak and nitrite peaks overlap with conductivity detectors. UV detects nitrite and nitrate, but not chloride. Electrical conductivity detection is based on detecting the conductivity of ions. Voltage is applied to a pair of plates, and if ions are present, there is a current. If there are ions in the mobile phase as it passes through the detector flow-cell a signal is produced that is proportional to the ion concentration. Each ion also has an equivalent conductivity constant that indicates how easily it conducts current. There are two type of conductivity detection. Non-suppressed conductivity and suppressed conductivity. Most EPA approved methods use the suppressed conductivity method. Standard Methods 4110C is EPA approved for wastewater and uses non-suppressed conductivity.

The simplest form of ion chromatography is non-suppressed conductivity (Figure 49). In this method, the mobile phase is a non-UV absorbing weak organic acid. Detection limits are high enough that the method is useful for the higher concentrations of anions in wastewater. The detector is positioned immediately after the mobile phase emerges from the column. The conductivity detector detects ions in the mobile phase, and cations and anions in the sample. This technique enables you to do ion chromatography using a normal HPLC.

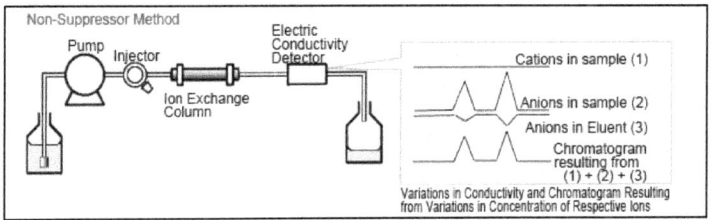

Figure 49, schematic of non-suppressed conductivity ion chromatography[16]

16

http://www.shimadzu.com/an/hplc/support/lib/lctalk/64intro.html,accessed February 24, 2017

In non-suppressed conductivity, each ion passes through the column to the detector. As mentioned before, the technique is not that sensitive for anions making it useful for wastewater. Cations sensitivity is not significantly affected by suppressed or non-suppressed conductivity. Most IC manufactures will tell you "don't bother with non-suppressed conductivity methods". However, EPA has approved a non-suppressed method for anions in wastewater. If you are tired of diluting samples, just to put chloride and sulfate on scale this may be an option.

The suppressor method (Figure 50) changes the eluent composition to one with a lower electrical conductivity. Solutions with aqueous sodium carbonate or sodium hydroxide are used. The suppressor is attached between the column outlet and the detector. The suppressor exchanges all the cations for hydrogen ions. Because hydrogen ions have a higher equivalent electrical conductivity the signal is increased.

Thus, the sensitivity of suppressed conductivity is higher.

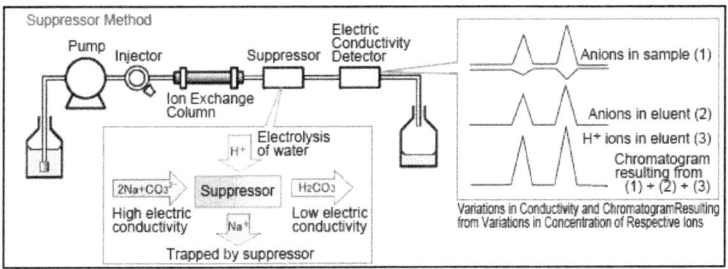

Figure 50, Schematic of suppressed conductivity ion chromatography[17]

The suppressor serves to reduce background noise and increase response. These IC methods have lower detection limits for anions than non-suppressed methods

---

17

http://www.shimadzu.com/an/hplc/support/lib/lctalk/64intro.html, accessed February 24, 2017

Suggestions on selection an ion chromatograph

Every IC, regardless of suppressed or non-suppressed detection has a pump, and injector, a column, a detector, and a data system. You really need an auto-sampler, so just get one.

These are things to ask yourself as you consider an IC:

1) Is the tubing made of PEEK?
2) Is it capable of pressures up to 5000 psi?
3) Does the conductivity detector detect up to 5000 micro-Siemens or higher?
4) Does the auto-sampler use plastic vials?
5) Is the column temperature controlled?
6) How much bench-space is required?

Most likely, you will need suppressed conductivity detection. The pump is important. If it pulses, you will have a noisy baseline. Get a 2-piston pump. Do the injector's fail? Is the detector low noise and high stability? Make sure the detector is heated. Do you have to replace them? How good is the service and support? How often do you have to make reagents? Will it make reagents for you?

Table 17 lists EPA approved methods for anions in wastewater

Table 17, EPA approved methods for anions in wastewater

Method	Analyte	detection
EPA 300.1	anions	Suppressed conductivity
SM 4110C	anions	Non-suppressed conductivity

High Performance Liquid Chromatography (HPLC)

High performance liquid chromatography is used for separating and analyzing organics in a mixture. It has an advantage over gas chromatography methods because you can run compounds in water samples with very little sample preparation. HPLC has been used in laboratories for over 30 years. A common environmental laboratory method is the analysis of carbamate pesticides. Recent advances in LC detection have added mass spectrometry and even more recent mass spec mass spec. These new detectors enable very fast, selective LC runs for polar pesticides and herbicides, cyanotoxins, perfluorinated compounds and pharmaceutical and personal care products. You can expect that the newer environmental methods will be LCMSMS.

In LC, the carrier is the mobile phase. The mobile phase is a single solvent, a mixture of solvents, or a variable mixture of solvents over time, called a gradient. The solvent selectively dissolved sample components injected and adsorbed onto a column to give you peaks. The mobile phase is pumped into very narrow bore tubing resulting in very high pressures. The narrower the tubing, the higher the pressure. Dissolved air in the mobile phase can lead to noise and drift in the baseline. Ultrasonic degassing with vacuum gives the best results. Helium sparging replaces air in the reagent with inert gas. Ultrasonic degassing creates nucleation sites where bubbles can form and escape from liquid.

To reconfirm the importance of degassing mobile phases, Figure 51 shows an example of problems that can occur due to air in the mobile phase. These problems can be resolved through appropriate degassing.

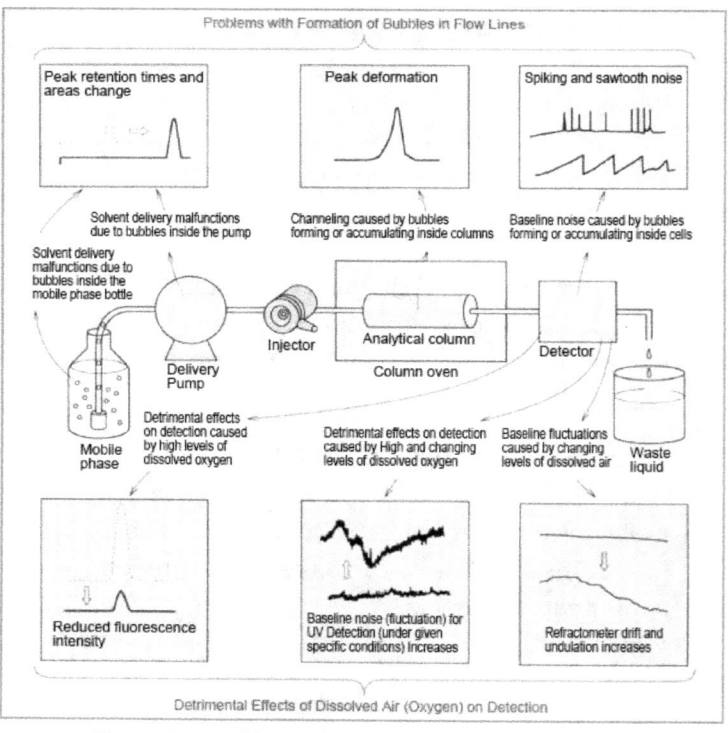

Figure 51, problems that can occur in the HPLC mobile phase.

The purpose of the pump in an HPLC is to provide a constant, reproducible supply of mobile phase to the column. The mobile phase enters a mixer. If it is not properly mixed, you will not get reproducible separation and elution from the column. A mixer is required in all instances when more than one solvent is used as mobile phase. This is to ensure sufficient mixing of all compounds before being introduced to the column. Factors to consider when choosing a mixer: flow rate, solvents used, and type of detector. The mixed mobile phase passes through a valve. It is at this valve where sample is loaded into a loop. The valve turns and the mobile phase carries the sample into the column. A HPLC injector is a high-pressure valve. You fill a fixed volume loop. This is contrary to GC methods where you are injecting sample through a septa into a carrier gas stream. Both manual injectors and auto-samplers use 6-port rotary valves to aspirate (load) and dispel (inject) sample. Gradients

(Figure 52) are very useful in adjusting retention and sharpening peaks (practical efficiency).

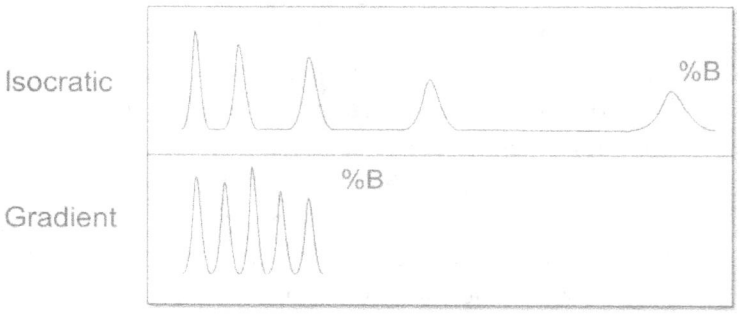

Figure 52, advantage of gradient systems in throughput and less band broadening

Depending on the analyte and the composition of the mobile phase, the analyte elutes from the column to pass through a detector flow cell. The column is specially made to retain analyte. The analyte elutes in very reproducible order at very reproducible times. The elution time determines the analyte identity. The intensity at the detector is a function of concentration. Once the solution passes through the detector it either travels to another detector in a series, or goes to waste.

Always stay within manufacturer's pH recommendations. Keep the temperature of the column less than 80°C or you will shorten its life. Record the pressure when you install the column and monitor it. As the column plugs the pressure will increase. Keep a record of column use – how many injections on a column. If it isn't designated for a single method – record conditions The first time you use a column, keep a baseline of pressure and initial QC run of HPLC column, either with test mix we used or better yet standards or analytes that you will be working with.

Rinse the column to "clean" column - prevent build up. To clean sample and matrix use a strong solvent. Cleaning will decrease the background signal, carry-over peaks, and extend the lifetime of the column.

Rinse buffers from columns before storage. (9+1) water: Organic solvent is good for rinsing. Use an appropriate shutdown method to extend column life and ensure column will sit overnight in the appropriate solvent.

To get rid of chromatographically retained analytes and particulates (particulate matter); flip column around and flush; a column filter will help prevent particulates in the column.

To get rid of chromatographically retained undesirable debris (dissolved matter that sticks to the phase of the column), change mobile phase composition to elute off or use a guard column.

A column oven provides a temperature-controlled environment during the separation process. The oven also plays an important role in controlling the quality and time for analysis (Figure 53).

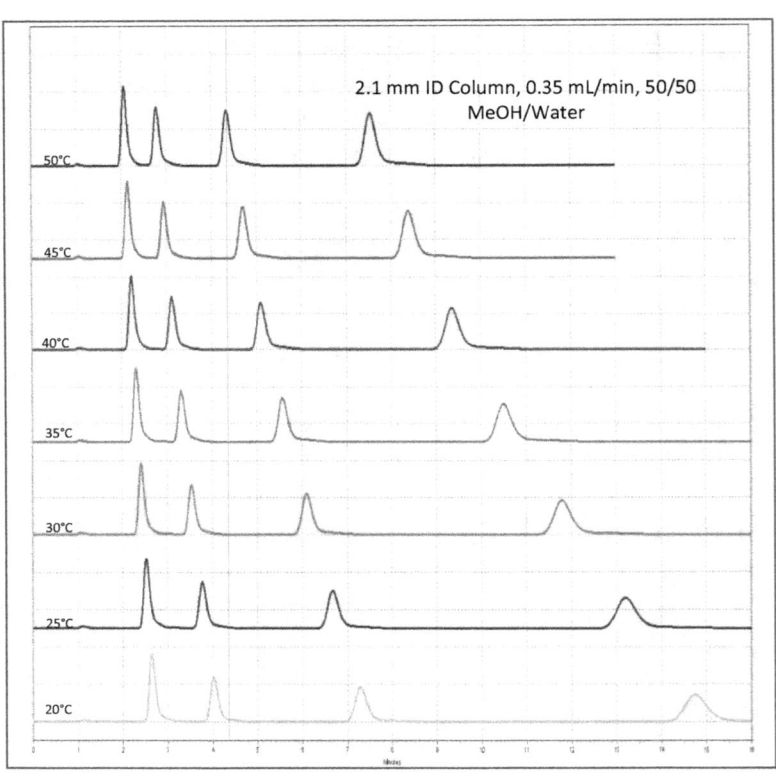

Figure 53, difference in chromatography with temperature

The detector continuously monitors the baseline as it passes through a flow-cell. Detectors measure some characteristic of the analytes you are looking for, and hopefully do not measure non-analytes in the matrix. The type of detector varies according to the method. In EPA methods, there are only a few HPLC detectors commonly used.

The most common LC detectors for environmental analysis are UV absorbance, fluorescence, and mass spectrometry. A UV detector can be used to measure PAHs. Essentially, any compound that absorbs light at the selected wavelength is measured. A fluorescence detector is used to measure carbamates after a post column derivatization. It can also be used as a more sensitive detection of PAH. Newer EPA methods are being developed using LCMSMS. This new technique requires little or no sample preparation. We may finally see the day when environmental testing can be done by direct injection. The LCMSMS technique is similar to GCMSMS previously described.

Liquid Chromatography Triple Quadrupole Mass Spectrometry (LCMSMS)

Unlike gas chromatography that is limited to volatile compounds, or those that can be extracted into an organic solvent and vaporized in a GC, an LCMSMS can analyze just about anything that happens to be dissolved in your sample. It is especially useful for those compounds that are difficult by GC methods, such as halo acetic acids, 2, 4-D and 2, 4, 5 -TP. It can take away the necessity of post column derivatization and analyze carbamates. Better yet, it can do all of this in a single sample run with little or no sample preparation and in very short times. Because the mass of product ions is so specific, co-elution is not as big a deal anymore. Even if you have multiple components eluting at the same time, the LCMSMS can still identify and quantify them based on mass.

As we described in the GCMSMS portion, a compound is ionized and passes through one quadrupole that selects one mass (Figure 54). Then that mass enters a collision cell where it is fragmented into other highly reproducible compounds of different mass. Another quadrupole selectively allows single mass ions to pass through and measures them on a detector.

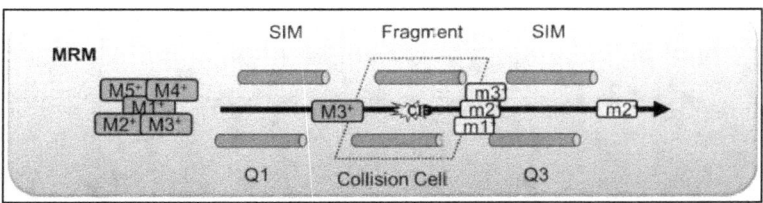

Figure 54, schematic of LCMSMS

Unlike GCMS, there is no real fragmentation pattern from the first ionization. Figure 55 compares the GCMS and LCMS spectra of riboflavin.

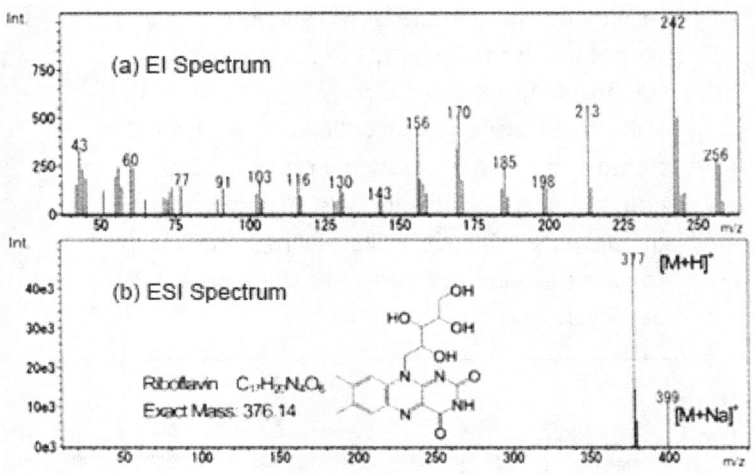

Figure 55, difference in GCMS (EI) and LCMS (ESI) fragmentation spectra

However, like GCMS, after the molecular ion passes through the collision cell it does exhibit a highly reproducible fragmentation pattern. Figure 56 shows that MRM fragmentation patterns allow a building of a database enabling library searches to identify unknowns.

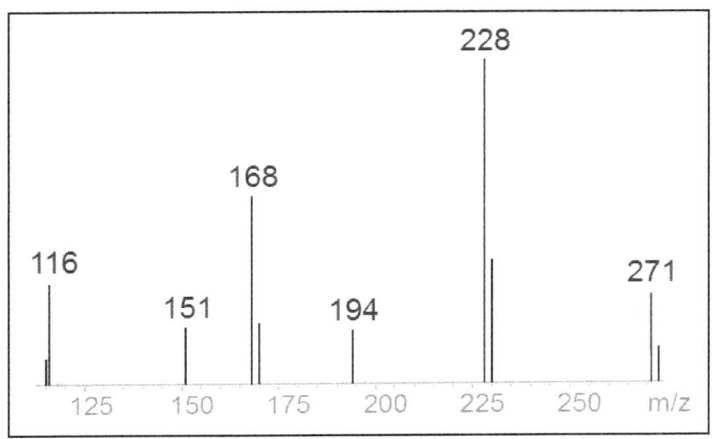

Figure 56 Collision induced spectra of multiple MRM

Modern instruments enable measurement of multiple fragments from one sample. These multiple fragments serve as confirmation that the identity of the peak is what you think it is. In Figure 57, eleven fragments were measured from one sample component.

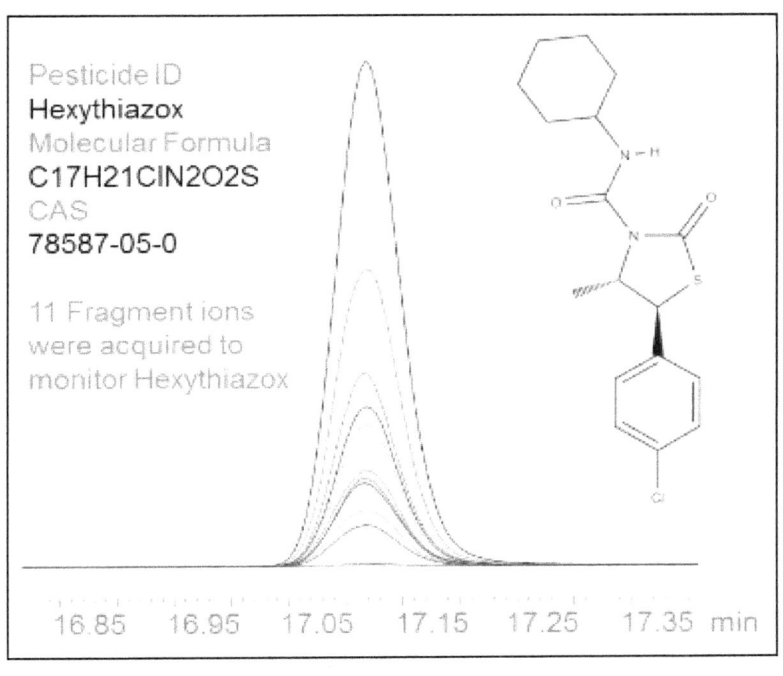

Figure 57, eleven MRM fragments

Each MRM fragment can be quantitated providing quantitative data from the eleven individual masses from the same compound.

Fast LCMS detectors enable very fast run times. In Figure 58, at the top you see slow data acquisition and the inability of the detector to resolve the numerous peaks eluting in the window.

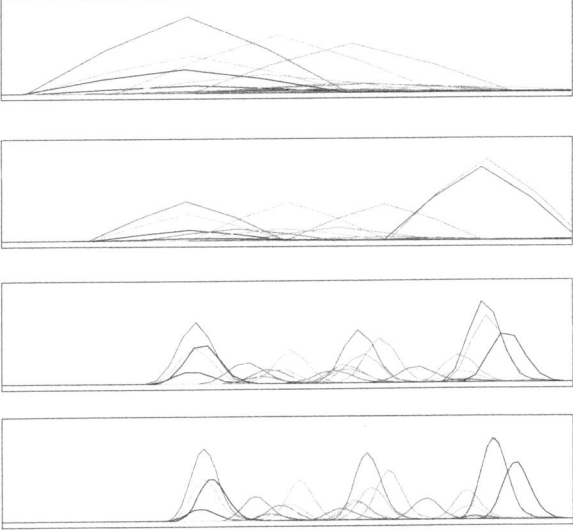

Figure 58 difference between slow (top) and fast (bottom) scanning

Faster scanning means more points per second, which literally pulls peaks out of the noise. At the bottom, you can see all the peaks that eluted within the window. In Figure 59 are twenty-five pesticide compounds all eluting between 6.3 and 6.7 minutes.

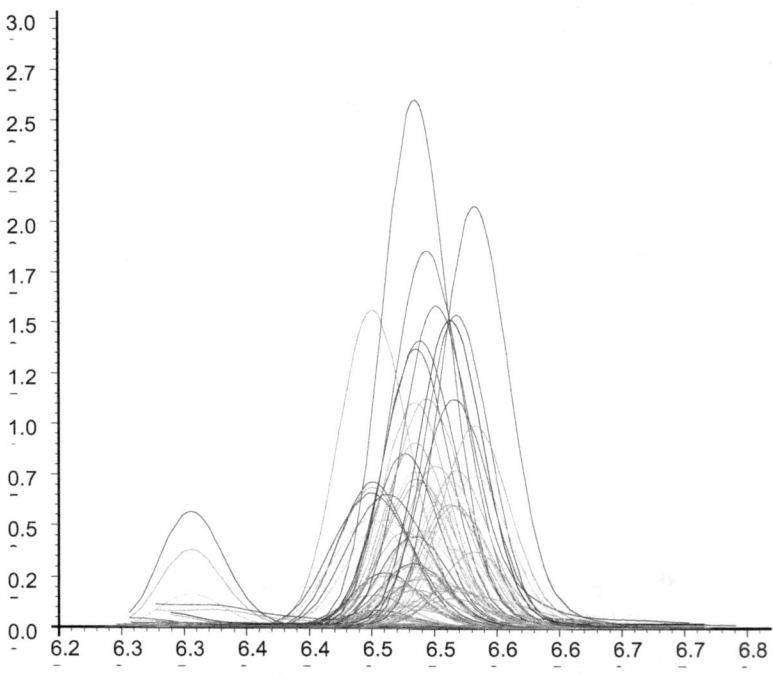

Figure 59 twenty-five pesticides eluting between 6.3 and 6.7 minutes

In addition, you want rapid polarity switching. In LCMSMS, you can measure both positive and negative ions. If you cannot switch polarity fast enough, you would have to measure positive ions in one run, then negative ions in another. With rapid polarity - switching you can measure both in a single run.

What Are the Advantages of LC-MS?

In general, liquid chromatography (LC) separates the components of a sample based on differences in their affinity for the stationary phase or mobile phase, and then detects the separated components using UV, fluorescence, or electrical conductivity detectors. Identification is on retention time and quantitation uses peak intensity or peak area. Chromatography may offer great resolution, but accurately qualifying and quantitating substances can is difficult when multiple components elute approximately at the same time. In contrast, mass spectrometry (MS) offers a highly sensitive detection technique that ionizes the sample components, then separates the resulting ions based on their mass-to-charge ratios and measures the intensity of each ion. Therefore, LC-MS systems combine the separation resolution of liquid chromatography with the qualitative capabilities of mass spectrometry. The mass spectra obtained from these scan measurements provides molecular mass and structural information for eluted components, which supplements the qualitative information based on retention times obtained using other LC detectors.

Suggestions for selecting an LCMS, or LCMSMS
1) Scan speed; the higher the scan speed the shorter you can make your run times.
2) Rapid polarity switching. You may be able to run both positive and negative ions in one injection saving you time.
3) Sensitivity. Current LCMS and LCMSMS methods extract by Solid Phase Extraction (SPE), concentrate the extract volume, and inject a few microliters of extract. Newer, highly sensitive LCMSMS instruments may

be capable of direct injection of sample mixed with organic solvent. This could eliminate the extraction step, boosting productivity and minimizing contamination and loss of analyte.
4) Maintenance. How hard is it to replace consumables or to clean? This is particular important for methods that direct inject.
5) Does the same manufacturer make the LC and the MS? This is important for between instrument communication, technical support, and service.
6) How good is the service and technical support? LCMSMS is new to environmental testing. Does the company understand environmental?

Summary

In summary, remember that the main specification for any instrument purchase is that it is capable of running the method that you are buying it to do. Ask the vendor for data proving that it can. Get detection limits on real samples if possible, however, you may need to overlook detection limits on samples that require digestions (these detection limits are lab specific). Avoid "locking out" vendors simply because your experience is with another brand. This preference may prevent you from getting better detection limits, better service, or more up time. Compare on-line, if possible, but also go to tradeshows and talk to the vendors.

Establish a good relationship with your sales or technical support people. Choose the instrument you need now and for the future. No more and no less. Your sales or technical support person may be able to provide insight as to new regulations coming. Remember that service and technical is a huge part of the cost or value of your instrument. It is very costly if you do not have it and provides unquantifiable value if you do. The hidden value of good technical support cannot be overrated. Remember the skill level of your employees now, and in the future.

About the Author

William is currently the general manager of Government and Regulatory Business Development at Shimadzu Scientific Instruments, Inc. Columbia, Maryland. William is the ASTM D19 committee on water Chair, the Standard Methods for the Examination of Water and Wastewater Part 4000 coordinator and AWWA editor, and an ISO TC 147 (water) SC2 (chemistry methods) delegate and expert. Prior to joining Shimadzu in 2014, William developed a new commercially available on-line cyanide analyzer for determination of cyanide in gold process solutions, its sample filtration module, redesigned existing continuous flow and discrete analyzers, designed a new Flow Injection Analyzer, and developed and validated numerous methods. William's work on sampling, preservation, and minimization of interferences in cyanide analysis led to the development of six new or modified ASTM standards, five of which are EPA approved at 40 CFR Part 136.3. William also authored, or co-authored methods for nitrate, ammonia, TOC, and TKN that are also approved at 40 CFR Part 136.3 and collaborated with EPA in defining method flexibility at 40 CFR Part 136.6. William wrote an ASTM guide to the International Cyanide Management Code and was a contributing author to the SME "Management Technologies for Metal Mining Influenced Water, Volume 6: Sampling and Monitoring for the Mine Life Cycle". William continuously works on new methods intended for EPA approval, including ASTM D8083 Total Nitrogen, pesticides and PCBs, Perfluorinated compounds, and nitrosamines. Besides experience in manufacturing, William has worked at several commercial testing laboratories as owner, Chief Science Officer, Chief Chemist, manager, bench chemist, and environmental consultant.

In addition to work at consensus organizations, William Lipps has also given over 100 podium and poster **presentations at** national and local laboratory and mining association conferences. William has given numerous webinars and has taught several short courses including topics such as "cyanide analysis", "wet chemistry analysis" and "how to select laboratory instruments". William has authored numerous technical guides and application notes for OI Analytical and Shimadzu and is a contributing author to LCGC and American Laboratory.

www.ingramcontent.com/pod-product-compliance
Lightning Source LLC
Chambersburg PA
CBHW060835220526
45466CB00003B/1117